Special Topics in
Electromagnetics

Special Topics in
Electromagnetics

$$\nabla \times \vec{E} = -\frac{\partial \vec{B}}{\partial t}$$

$$\nabla \times \vec{H} = \vec{J} + \frac{\partial \vec{D}}{\partial t}$$

$$\nabla \cdot \vec{D} = \rho$$

$$\nabla \cdot \vec{B} = 0$$

Kun-Mu Chen
Michigan State University, USA

臺大出版中心
NATIONAL TAIWAN UNIVERSITY PRESS

Published by

National Taiwan University Press
1, Sec 4, Roosevelt Road
Taipei, Taiwan

and

Distributed by

World Scientific Publishing Co. Pte. Ltd.

5 Toh Tuck Link, Singapore 596224

USA office: 27 Warren Street, Suite 401-402, Hackensack, NJ 07601

UK office: 57 Shelton Street, Covent Garden, London WC2H 9HE

British Library Cataloguing-in-Publication Data
A catalogue record for this book is available from the British Library.

SPECIAL TOPICS IN ELECTROMAGNETICS

ISBN 978-981-4412-17-9

Printed in Singapore

To my wife June and our children

Maggie, Kathy, Kenny and George

Preface

This book is the outgrowth of the lecture notes I developed while teaching at National Taiwan University as a visiting professor during the period of December 2006 to February 2007. Five Special topics in Electromagnetics are covered in this book which can be taught as a graduate course in Electromagnetics. The prerequisite to understand this book is some experience in undergraduate courses on this subject. The format of this book is somewhat unusual because the material is presented with detailed mathematical derivation, with an intention for self-taught students and researchers to read this book. The special topics covered in this book are becoming more popular in research areas of interaction of Electromagnetic field with material bodies, the practical applications of Electromagnetic wave in biomedical engineering, and the radar target identification.

Chapter 1 presents a set of modified Maxwell's equations with magnetic sources and the application of these equations in solving problems in heterogeneous multiple media. The Equivalence Principle is mathematically derived and its application to the interaction of an EM wave with a dielectic body is discussed.

Chapter 2 derives Vector Wave Functions and demonstrates their applications to the problems involving the interaction of EM waves with material bodies in various environments. The result of an analysis gives the answer to the question of why the sky looks blue.

Chapter 3 deals with Dyadic Green's functions and their applications. The convergences problem of Dyadic Green's functions when applied in some source regions or conducting regions is discussed in details. The principal value integration and the correction term are derived. These mathematical problems are explained with physical pictures to facilitate the understanding of Dyadic Green's functions.

Chapter 4 presents a practical application of EM waves in biomedical area of microwave life-detection system. The scheme of detecting a very small movement of biological body with EM waves is studied.

Chapter 5 presents a new method of the radar target identification with EM pulse and the scheme of the Extinction-pulse method.

The author is indebted to the College of Electrical Engineering and Computer Science, National Taiwan University, the author's Alma Mater, for their assistance. Thanks are expressed to Ms. Wening Chao, Mr. Yung-Chih Tsai, Mr. Yu-Che Lin, Mr. Jing-De Huang, Mr. Guo-Cyuan Guo and Ms. Hui-Jun Lian for typing the first draft of this book. The hospitality rendered by Prof. Ruei-Beei Wu during my stay at National Taiwan University is appreciated. Furthermore, my sincere gratitude is expressed to Prof. Huei Wang, my former student now a famous professor himself, for his many helps in arranging my visit to Taiwan and the preparation of this book. Without his help this book will not be possible.

I should not forget to acknowledge the contribution made by many of my former students and Prof. Dennis P. Nyquist and Prof. Edward J. Rothwell of Michigan State University, East Lansing, Michigan, where I taught for about forty years. Many results presented in this book were produced by them under my supervision.

Kun-Mu Chen

San Diego, California. 2007

CONTENTS

Preface ... *vii*

Chapter 1. *General Maxwell's Equations and Solutions* ... *1*

1.a Modified Maxwell's Equations with Magnetic Source Terms ... **1**

1.b General Solutions of Maxwell's Equations in Heterogeneous Multiple Media ... **6**

1.c Mathematical Derivation of Equivalence Principle ... **28**

1.d Applications ... **34**

Chapter 2. *Vector Wave Functions and Applications* ... *39*

2.a Wave Equations ... **39**

2.b Spherical Scalar Wave Functions ... **41**

2.c Spherical Vector Wave Functions ... **46**

2.d Scattering of a Conducting Sphere — Application ... **54**

2.e Interaction of a Material Body with EM field in a Rectangular Cavity ... **72**

Chapter 3. *Dyadic Green's Functions and Applications* *87*

3.a **Dyadic Analysis** 87

3.b **Dyadic Green's Function** 91

3.c **Dyadic Green's Function in Source Region or Conducting Medium** 97

3.d **Applications of Dyadic Green's Function** 105

3.e **Physical picture of Dyadic Green's Function in Source Region** 109

Chapter 4. *Biomedical Application of Electromagnetic Waves* *117*

4.a **Microwave Life-Detection Systems** 118

4.b **R.F. Life-Detection Systems for Searching Human Subjects under Earthquake Rubble or Behind Barrier** 133

4.c **Analysis of Interaction Between ELF-LF Electric Fields and Human Bodies** 153

Chapter 5. *Radar Target Identification with Extinction-pulse (E-pulse) Method* *177*

5.a **Radar Target Identification by E-pulse Method — Time-Domain Analysis** 178

5.b **E-pulse Method — Frequency Domain Analysis** 195

5.c **Application of E-pulse Technique to Early-Time Target Response** 205

Index *211*

Chapter 1
General Maxwell's Equations and Solutions

1.a Modified Maxwell's Equations with Magnetic Source Terms

Conventional Maxwell's equations which express the relations between the field vectors, the electric field $\vec{E}$ and the magnetic induction $\vec{B}$, and the source current density $\vec{J}$ and charge density ρ are usually written in the following form :

$$\begin{cases} \nabla\times\vec{E}=-\dfrac{\partial\vec{B}}{\partial t} & (1.1) \\ \nabla\times\vec{H}=\vec{J}+\dfrac{\partial\vec{D}}{\partial t} & (1.2) \\ \nabla\cdot\vec{D}=\rho & (1.3) \\ \nabla\cdot\vec{B}=0 & (1.4) \end{cases}$$

where the displacement vector $\vec{D}$ and the magnetic field $\vec{H}$ are defined by

$$\begin{cases} \vec{D}=\varepsilon_0\vec{E}+\vec{P} & (1.5) \\ \vec{H}=\vec{B}/\mu_0-\vec{M} & (1.6) \end{cases}$$

with $\vec{P}$ as the polarization density, $\vec{M}$ as the magnetization density, and ε_0 and μ_0 are permittivity and permeability of the free space.

The source current density $\vec{J}$ and source charge density ρ are related by the equation of continuity:

$$\nabla\cdot\vec{J}+\frac{\partial\rho}{\partial t}=0 \qquad (1.7)$$

In a simple medium,

$$\begin{cases} \vec{D} = \varepsilon \vec{E} & (1.8) \\ \vec{H} = \vec{B}/\mu & (1.9) \end{cases}$$

Where ε and μ are the permittivity and permeability of the medium.

The Maxwell's equations expressed in Eqs. (1.1) to (1.4) are not symmetrical mathematically. Equations (1.1) and (1.2) are different in form and so are Eqs. (1.3) and (1.4) as the result. Why? Because the elementary source for the electric field $\vec{E}$ is different from that of the magnetic field $\vec{B}$. Eq. (1.3) implies that $\vec{E}$ field starts from or terminates at an electric charge ρ. However, Eq. (1.4) implies that $\vec{B}$ field does not start from or terminate at a magnetic charge. This is due to the fact that there exists no real magnetic charge (or magnetic monopole). In fact, the smallest unit of the magnetic source is a magnetic dipole which is actually an infinitesimal current loop. The situation is depicted in Fig. 1a. This is the conventional thinking even though there are some reports on the discovery of magnetic monopoles. If these reports are verified in the future, it would not alter the following derivation of the modified set of Maxwell's equations with magnetic source terms.

Now, let's get back to the conventional thinking. Let's consider a general case where the source current consists of two types of currents, namely, a linear current which flows or oscillates in linear directions and an infinitesimal circulatory current which flows or oscillates around an infinitesimal loop.

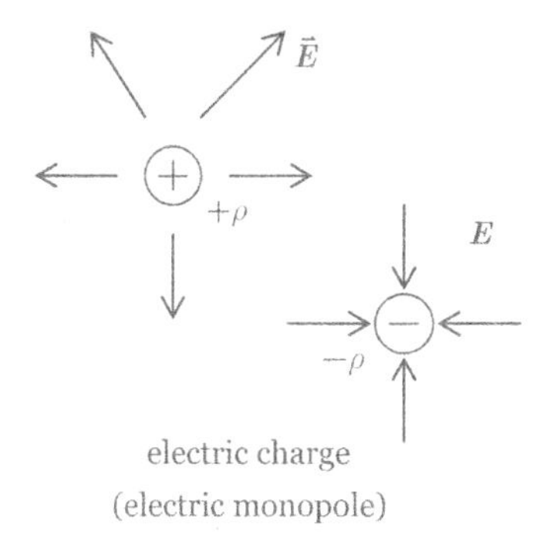

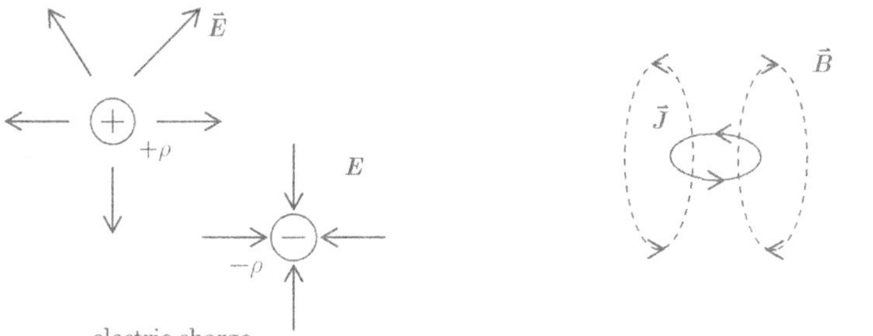

electric charge
(electric monopole)

infinitesimal current loop

Fig. 1a

electric dipole
(steady case)

equivalent magnetic dipole
(steady case)

Fig. 1b

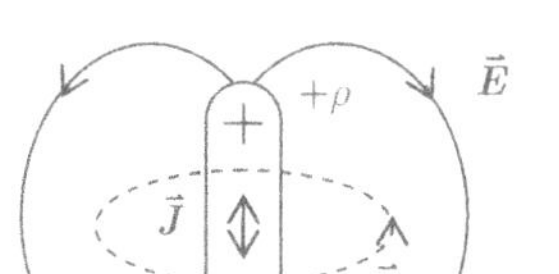

electric dipole
(dynamic case)

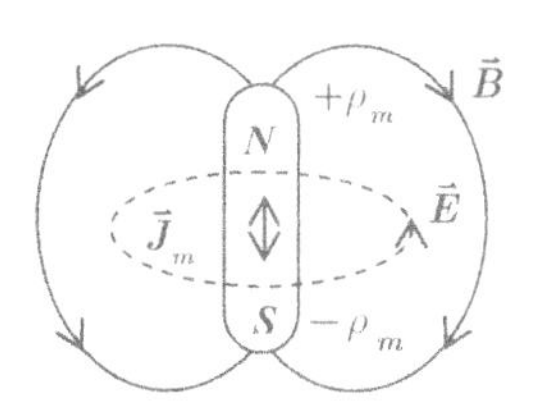

magnetic dipole
(dynamic case)

Fig. 1c

For this general case, it is rather hard to mathematically describe $\vec{J}$ which include both types of currents. Mathematically it is sometimes advantageous to recognize these linear currents as $\vec{J}_e$ and imagine those infinitesimal circulatory currents as equivalent magnetic dipoles, because a magnetic dipole produces the same magnetic field as an infinitesimal current loop and it is easier to express a magnetic dipole mathematically, analogously as an electric dipole source. The situation is depicted in Fig. 1b. If the circulatory current varies its magnitude and direction or oscillates around the loop with time, the equivalent magnetic dipole strength oscillates with time likewise. When the magnetic dipole oscillates with time, it can be considered as an imaginary magnetic current flowing up and down. This is the complete analogy of a time-varying electric dipole, as depicted in Fig. 1c. Therefore, a time-varying circulatory current can be represented by equivalent magnetic current and charge as $\vec{J}_m$ and ρ_m.

Let's imagine now that the magnetic charge ρ_m produces $\vec{B}$ field in an analogous way as the electric charge ρ_e produces $\vec{E}$ field. Then Eq. (1.4) should be modified as

$$\nabla \cdot \vec{B} = \rho_m \tag{1.10}$$

If Eq. (1.4) is modified to become Eq. (1.10), Eq. (1.1) also needs modification. Because

$$\nabla \cdot (\nabla \times \vec{E}) = 0 = -\nabla \cdot \left(\frac{\partial \vec{B}}{\partial t} \right) = -\frac{\partial}{\partial t}(\nabla \cdot \vec{B}) = -\frac{\partial}{\partial t}\rho_m$$

which is not true since ρ_m is a function of time.

If we assume that relation between ρ_m and $\vec{J}_m$ follow an usual continuity equation such as Eq. (1.7), we should have

$$\nabla \cdot \vec{J}_m + \frac{\partial \rho_m}{\partial t} = 0 \tag{1.11}$$

With Eqs. (1.11) and (1.10), Eq. (1.1) should be modified to

$$\nabla \times \vec{E} = -\vec{J}_m - \frac{\partial \vec{B}}{\partial t} \tag{1.12}$$

With these modifications, Maxwell's equations can now be written as

$$\begin{cases} \nabla \times \vec{E} = -\vec{J}_m - \dfrac{\partial \vec{B}}{\partial t} & (1.13) \\ \nabla \times \vec{H} = \vec{J}_e + \dfrac{\partial \vec{D}}{\partial t} & (1.14) \\ \nabla \cdot \vec{D} = \rho_e & (1.15) \\ \nabla \cdot \vec{B} = \rho_m & (1.16) \end{cases}$$

Equations of continuity for the current and charge for electric and magnetic sources are

$$\nabla \cdot \vec{J}_e + \frac{\partial \rho_e}{\partial t} = 0 \tag{1.17}$$

$$\nabla \cdot \vec{J}_m + \frac{\partial \rho_m}{\partial t} = 0 \tag{1.18}$$

With the addition of ρ_m and $\vec{J}_m$, Maxwell's equations appear to be more symmetrical mathematically.

However, it is important to know that this magnetic source is an equivalent source of infinitesimal current loops, and it does not bear sound physical meanings. Also, the equivalent magnetic source model may not be justified at microscopic scale. It is important to remember that if the circulatory source currents are described mathematically as a part of $\vec{J}$, the original set of Maxwell's equations given in Eqs. (1.1) to (1.4) is sufficient to describe the relations between the field vectors and the sources.

It is noted that if the real magnetic charges or magnetic monopoles are discovered and verified in the future, the magnetic current and charge, $\vec{J}_m$ and ρ_m, should appear in Eqs. (1.13) and (1.16) as shown. They will also maintain $\vec{E}$ and $\vec{B}$ fields.

1.b General Solutions of Maxwell's Equations in Heterogeneous Multiple Media

We desire to find the field vectors, the $\vec{E}$ and $\vec{H}$ fields, maintained by electric and magnetic sources located inside two dissimilar regions. This is not a simple case of a homogenous infinite region and Green's functions for an infinite medium do not apply. To avoid complexity in notation, we will use $(\vec{J}, \rho)$ as electric current and charge source densities, and $(\vec{M}, \rho_m)$ as magnetic current and charge source densities. The two regions are simple media which can be characterized by their respective permittivity ε and permeability μ. Also we will consider the time-harmonic case or the Fourier-transformed forms of general Maxwell's equations given in the preceding section.

The geometry under consideration is depicted in Fig. 2. This geometry consists of region 2 with complex permittivity and permeability (ε_2, μ_2), the volume V_2, the boundary surface S, and the electric and magnetic source currents $(\vec{J}_2, \vec{M}_2)$ within V_2. Region 2 is surrounded by region 1 of infinite volume V_1 that has electric parameters of (ε_1, μ_1) and source currents of $(\vec{J}_1, \vec{M}_1)$ within V_1.

We aim to find the EM fields in regions 1 and 2 in terms of the given source currents and equivalent surface currents on S.

Maxwell's equations for region 1 and 2 are

$$\begin{cases} \nabla\times\vec{E}_1 = -\vec{M}_1 - j\omega\mu_1\vec{H}_1 \\ \nabla\times\vec{H}_1 = \vec{J}_1 + j\omega\varepsilon_1\vec{E}_1 \end{cases} \quad \text{in } V_1 \tag{1.19}$$

$$\begin{cases} \nabla\times\vec{E}_2 = -\vec{M}_2 - j\omega\mu_2\vec{H}_2 \\ \nabla\times\vec{H}_2 = \vec{J}_2 + j\omega\varepsilon_2\vec{E}_2 \end{cases} \quad \text{in } V_2 \tag{1.20}$$

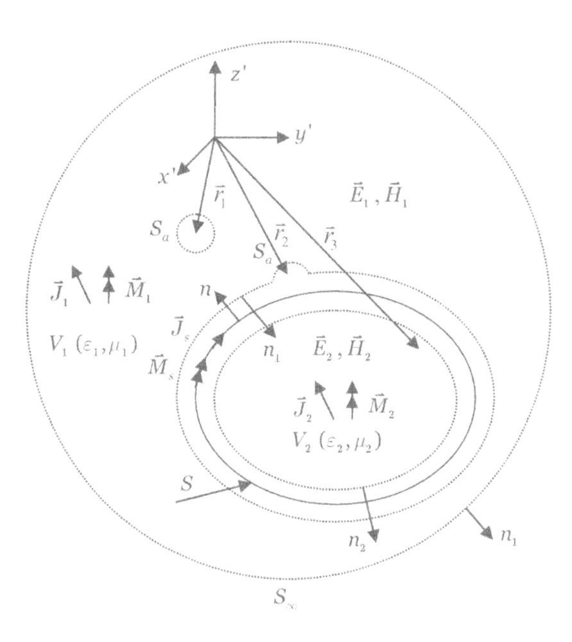

Fig. 2 Geometry of the problem: region 2 with volume V_2, boundary surface S, electric parameters (ε_2, μ_2), and source currents $(\vec{J}_2, \vec{M}_2)$ is surrounded by region 1 with infinite volume V_1, electric parameters (ε_1, μ_1), and source currents $(\vec{J}_1, \vec{M}_1)$. $(\vec{E}_1, \vec{H}_1)$ constitute the EM field in V_1 and $(\vec{E}_2, \vec{H}_2)$ that in V_2.

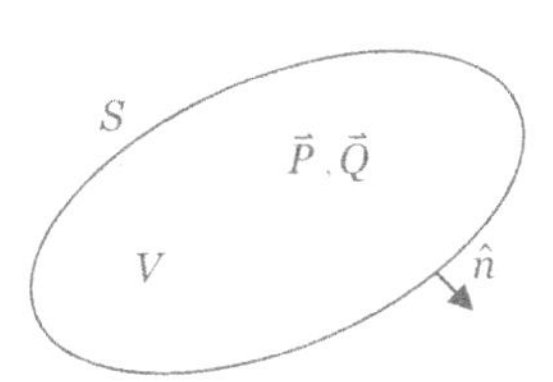

Fig. 3 A volume V surrounded by a boundary surface S with two vector functions $\vec{P}$ and $\vec{Q}$ inside.

Equations (1.19) can lead to wave equations for $\vec{E}_1$ and $\vec{H}_1$ as follows:

$$\begin{aligned}\nabla\times\nabla\times\vec{E}_1 &= -\nabla\times\vec{M}_1 - j\omega\mu_1\nabla\times\vec{H}_1 \\ &= -\nabla\times\vec{M}_1 - j\omega\mu_1(\vec{J}_1 + j\omega\varepsilon_1\vec{E}_1) \\ &= -\nabla\times\vec{M}_1 - j\omega\mu_1\vec{J}_1 + \omega^2\mu_1\varepsilon_1\vec{E}_1\end{aligned}$$

or

$$\nabla\times\nabla\times\vec{E}_1 - \beta_1^2\vec{E}_1 = -j\omega\mu_1\vec{J}_1 - \nabla\times\vec{M}_1 \tag{1.21}$$

where $\beta_1^2 = \omega^2\mu_1\varepsilon_1$

Similarly :

$$\begin{aligned}\nabla\times\nabla\times\vec{H}_1 &= \nabla\times\vec{J}_1 + j\omega\varepsilon_1\nabla\times\vec{E}_1 \\ &= \nabla\times\vec{J}_1 + j\omega\varepsilon_1(-\vec{M}_1 - j\omega\mu_1\vec{H}_1) \\ &= \nabla\times\vec{J}_1 - j\omega\varepsilon_1\vec{M}_1 + \omega^2\mu_1\varepsilon_1\vec{H}_1\end{aligned}$$

or

$$\nabla\times\nabla\times\vec{H}_1 - \beta_1^2\vec{H}_1 = -j\omega\varepsilon_1\vec{M}_1 + \nabla\times\vec{J}_1 \tag{1.22}$$

From Eq. (1.20) we will have

$$\nabla\times\nabla\times\vec{E}_2 - \beta_2^2\vec{E}_2 = -j\omega\mu_2\vec{J}_2 - \nabla\times\vec{M}_2 \tag{1.23}$$

and

$$\nabla\times\nabla\times\vec{H}_2 - \beta_2^2\vec{H}_2 = -j\omega\varepsilon_2\vec{M}_2 + \nabla\times\vec{J}_2 \tag{1.24}$$

where $\beta_2^2 = \omega^2\mu_2\varepsilon_2$

In the derivation of $\vec{E}(\vec{r})$ and $\vec{H}(\vec{r})$, we need the vector Green's theorem such as

$$\int_V(\vec{Q}\cdot\nabla\times\nabla\times\vec{P} - \vec{P}\cdot\nabla\times\nabla\times\vec{Q})dv = \int_S(\vec{P}\times\nabla\times\vec{Q} - \vec{Q}\times\nabla\times\vec{P})\cdot\hat{n}\,ds \tag{1.25}$$

This theorem can be proved simply as follow : Consider a finite volume V surrounded by a boundary surface S as shown in Fig. 3. $\vec{P}$ and $\vec{Q}$ are two vector functions which are continuous and their first and second derivations are also continuous throughout V. $\hat{n}$ is the unit normal vector pointing outward from V.

Consider a vector function, $\vec{P}\times\nabla\times\vec{Q}$,

$$\int_V \nabla\cdot(\vec{P}\times\nabla\times\vec{Q})dv = \int_S (\vec{P}\times\nabla\times\vec{Q})\cdot\hat{n}\,ds$$

where

$$\nabla\cdot(\vec{P}\times\nabla\times\vec{Q}) = (\nabla\times\vec{P})\cdot(\nabla\times\vec{Q}) - \vec{P}\cdot(\nabla\times\nabla\times\vec{Q})$$
$$\because\quad \nabla\cdot(\vec{a}\times\vec{b}) = \vec{b}\cdot\nabla\times\vec{a} - \vec{a}\cdot\nabla\times\vec{b}$$

so

$$\int_V [(\nabla\times\vec{P})\cdot(\nabla\times\vec{Q}) - \vec{P}\cdot(\nabla\times\nabla\times\vec{Q})]dv = \int_S (\vec{P}\times\nabla\times\vec{Q})\cdot\hat{n}\,ds \qquad (1.26)$$

Next, consider another vector function, $\vec{Q}\times\nabla\times\vec{P}$, we will get

$$\int_V [(\nabla\times\vec{Q})\cdot(\nabla\times\vec{P}) - \vec{Q}\cdot(\nabla\times\nabla\times\vec{P})]dv = \int_S (\vec{Q}\times\nabla\times\vec{P})\cdot\hat{n}\,ds \qquad (1.27)$$

Eq. (1.26) – Eq. (1.27) leads to

$$\int_V (\vec{Q}\cdot\nabla\times\nabla\times\vec{P} - \vec{P}\cdot\nabla\times\nabla\times\vec{Q})dv = \int_S (\vec{P}\times\nabla\times\vec{Q} - \vec{Q}\times\nabla\times\vec{P})\cdot\hat{n}\,ds \qquad (1.28)$$

Now, let us consider region 1 first and apply the vector Green's theorem of Eq. (1.28) to V_1. We will switch to the primed coordinates to have $\vec{r}\,'$ as an arbitrary source (integrating) point and $\vec{r}$ as a designated field (observation) point.

We choose

$$\vec{P} = \vec{E}_1(\vec{r}\,') \qquad (1.29)$$

and

$$\vec{Q} = \hat{a}\phi_1(\vec{r}\,',\vec{r}) = \hat{a}e^{-j\beta_1 R}/R = \hat{a}\,\exp(-j\beta_1|\vec{r}\,'-\vec{r}|)/|\vec{r}\,'-\vec{r}| \qquad (1.30)$$

$\vec{E}_1(\vec{r}\,')$ is the electric field at $\vec{r}\,'$ within V_1, $\hat{a}$ is a constant unit vector, and ϕ_1 is the unbounded Green's function for region 1. It is noted that if $\vec{r}$ is within V_1, $\vec{Q}$ will not be continuous at $\vec{r}\,'=\vec{r}$ and it is necessary to remove this singularity point before Eq. (1.28) can be applied.

When the field point $\vec{r}$ is an interior point within V_1, such as $\vec{r}_1$ in Fig 2, $\phi_1 \to \infty$ as $\vec{r}\,' \to \vec{r}_1$, thus we need to exclude this point with a small sphere having a small surface of S_a as depicted in Fig. 2. Then the total boundary surface S, for V_1 will consist of

$$S_1 = S + S_a + S_\infty$$

where S_∞ is the infinite spherical surface enclosing the outside of V_1.

The substitution of Eqs. (1.29) and (1.30) in Eq. (1.28) will lead to a lengthy manipulation as follows :

$$\nabla' \times \vec{Q} = \nabla' \times (\phi_1\, \hat{a}) = \nabla' \phi_1 \times \hat{a} + \phi_1 \underbrace{\nabla' \times \hat{a}}_{0} = \nabla' \phi_1 \times \hat{a}$$

$$\begin{aligned} \nabla' \times (\nabla' \times \vec{Q}) &= \nabla' \times (\nabla' \phi_1 \times \hat{a}) \\ &= (\nabla' \phi_1) \underbrace{(\nabla' \cdot \hat{a})}_{0} - \hat{a}\, \nabla' \cdot (\nabla' \phi_1) + (\hat{a} \cdot \nabla') \nabla' \phi_1 - (\nabla' \phi_1 \cdot \underbrace{\nabla') \hat{a}}_{0} \\ &= -\hat{a} \nabla'^2 \phi_1 + (\hat{a} \cdot \nabla') \nabla' \phi_1 \end{aligned}$$

$$\because \quad \nabla \times (\vec{a} \times \vec{b}) = \vec{a}\, \nabla \cdot \vec{b} - \vec{b}\, \nabla \cdot \vec{a} + (\vec{b} \cdot \nabla) \vec{a} - (\vec{a} \cdot \nabla) \vec{b}$$

also
$$\nabla'^2 \phi_1 + \beta_1^2 \phi_1 = -4\pi \delta(\vec{r}\,' - \vec{r}_1)$$

so for $\vec{r}\,' \neq \vec{r}_1$ ($\vec{r}_1$ is excluded by S_a)

$$\nabla'^2 \phi_1 = -\beta_1^2 \phi_1$$

moreover,

$$(\hat{a} \cdot \nabla') \nabla' \phi_1 = \nabla' (\hat{a} \cdot \nabla' \phi_1)$$

$$\because \quad \nabla' (\vec{a} \cdot \vec{b}) = (\vec{a} \cdot \nabla') \vec{b} + (\vec{b} \cdot \nabla') \vec{a} + \vec{a} \times (\nabla' \times \vec{b}) + \vec{b} \times (\nabla' \times \vec{a})$$

and
$$\nabla' (\hat{a} \cdot \nabla' \phi_1) = (\hat{a} \cdot \nabla') \nabla' \phi_1 + (\nabla' \phi_1 \cdot \underbrace{\nabla') \hat{a}}_{0} + \hat{a} \times (\underbrace{\nabla' \times \nabla'}_{0} \phi_1) + \nabla' \phi_1 \times (\underbrace{\nabla' \times \hat{a}}_{0})$$

Hence,

$$\nabla'\times(\nabla'\times\vec{Q})=\hat{a}\beta_1^2\phi_1+\nabla'(\hat{a}\cdot\nabla'\phi_1) \tag{1.31}$$

Now, for the other term in Eq. (1.28),

$$\nabla'\times\nabla'\times\vec{P}=\nabla'\times\nabla'\times\vec{E}_1=\beta_1^2\vec{E}_1-j\omega\mu_1\vec{J}_1-\nabla'\times\vec{M}_1 \tag{1.32}$$

based on Eq. (1.21).

substituting Eqs. (1.31) and (1.32) into Eq. (1.28),

$$\int_V[\phi_1\hat{a}\cdot(\beta_1^2\vec{E}_1-j\omega\mu_1\vec{J}_1-\nabla'\times\vec{M}_1)-\vec{E}_1\cdot(\hat{a}\beta_1^2\phi_1+\nabla'(\hat{a}\cdot\nabla'\phi_1))]dv'$$
$$=\int_{S_1}[\vec{E}_1\times(\nabla'\phi_1\times\hat{a})-\phi_1\hat{a}\times\nabla'\times\vec{E}_1]\cdot\hat{n}_1ds'$$
$$\because\ \nabla'\times(\hat{a}\phi_1)=\nabla'\phi_1\times\hat{a}+\phi_1\underbrace{\nabla'\times\hat{a}}_{0}$$

or

$$\int_{V_1}[-j\omega\mu_1\phi_1\hat{a}\cdot\vec{J}_1-\phi_1\hat{a}\cdot(\nabla'\times\vec{M}_1)-\vec{E}_1\cdot\nabla'(\hat{a}\cdot\nabla'\phi_1)]dv' \tag{1.33}$$
$$=\int_{S_1}[\vec{E}_1\times(\nabla'\phi_1\times\hat{a})-\phi_1\hat{a}\times\nabla'\times\vec{E}_1]\cdot\hat{n}_1ds'$$

where $S_1=S+S_a+S_\infty$

The left hand side of Eq. (1.33) can be simplified as follows :

Since

$$\nabla'\cdot[(\hat{a}\cdot\nabla'\phi_1)\vec{E}_1]=\nabla'(\hat{a}\cdot\nabla'\phi_1)\cdot\vec{E}_1+(\hat{a}\cdot\nabla'\phi_1)\nabla'\cdot\vec{E}_1$$
$$=\nabla'(\hat{a}\cdot\nabla'\phi_1)\cdot\vec{E}_1+(\hat{a}\cdot\nabla'\phi_1)\rho_1/\varepsilon_1$$

$$\vec{E}_1\cdot\nabla'(\hat{a}\cdot\nabla'\phi_1)=\nabla'\cdot[(\hat{a}\cdot\nabla'\phi_1)\vec{E}_1]-(\hat{a}\cdot\nabla'\phi_1)\rho_1/\varepsilon_1$$

Hence,

$$\begin{aligned}L.H.S. &= \int_{V_1}[-j\omega\mu_1\phi_1(\hat{a}\cdot\vec{J}_1)-\phi_1\hat{a}\cdot(\nabla'\times\vec{M}_1)+(\hat{a}\cdot\nabla'\phi_1)\rho_1/\varepsilon_1]dv' \\ &-\int_{V_1}\nabla'\cdot[(\hat{a}\cdot\nabla'\phi_1)\vec{E}_1]dv' \\ \because\ &\int_{V_1}\nabla'\cdot[(\hat{a}\cdot\nabla'\phi_1)\vec{E}_1]dv'=\int_{S_1}[(\hat{a}\cdot\nabla'\phi_1)\vec{E}_1]\cdot\hat{n}_1 ds' \\ &=\int_{S_1}(\hat{a}\cdot\nabla'\phi_1)(\vec{E}_1\cdot\hat{n}_1)ds'=\hat{a}\cdot\int_{S_1}(\nabla'\phi_1)(\hat{n}_1\cdot\vec{E}_1)ds' \\ &=\hat{a}\cdot\int_{S_1}(\hat{n}_1\cdot\vec{E}_1)(\nabla'\phi_1)ds'\end{aligned}$$

So

$$\begin{aligned}L.H.S. &= \hat{a}\cdot\int_{V_1}[-j\omega\mu_1\vec{J}_1\phi_1-(\nabla'\times\vec{M}_1)\phi_1+(\rho_1/\varepsilon_1)\nabla'\phi_1]dv' \\ &-\hat{a}\cdot\int_{S_1}(\hat{n}\cdot\vec{E}_1)(\nabla'\phi_1)ds'\end{aligned} \tag{1.34}$$

The right hand side of Eq. (1.33) can be rewritten as follows :

$$\begin{aligned}[\vec{E}_1\times(\nabla'\phi_1\times\hat{a})]\cdot\hat{n}_1 &= [(\vec{E}_1\cdot\hat{a})\nabla'\phi_1-(\vec{E}_1\cdot\nabla'\phi_1)\hat{a}]\cdot\vec{n}_1 \\ &=(\vec{a}\cdot\vec{E}_1)(\nabla'\phi_1\cdot\hat{n}_1)-(\hat{a}\cdot\hat{n}_1)(\vec{E}_1\cdot\nabla'\phi_1) \\ &=\vec{a}\cdot[\vec{E}_1(\nabla'\phi_1\cdot\hat{n}_1)-\hat{n}_1(\vec{E}_1\cdot\nabla'\phi_1)] \\ &=\vec{a}\cdot[\nabla'\phi_1\times(\vec{E}_1\times\hat{n}_1)]\end{aligned}$$

$$\because\ \vec{a}\times(\vec{b}\times\vec{c})=(\vec{a}\cdot\vec{c})\vec{b}-(\vec{a}\cdot\vec{b})\vec{c}$$

Also

$$\begin{aligned}[\phi_1\hat{a}\times\nabla'\times\vec{E}_1]\cdot\hat{n}_1 &= \phi_1[\hat{a}\cdot[(\nabla'\times\vec{E}_1)\times\hat{n}_1]] \\ &=\hat{a}\cdot[\phi_1(\nabla'\times\vec{E}_1)\times\hat{n}_1] \\ &=\hat{a}\cdot[\phi_1(-\vec{M}_1-j\omega\mu_1\vec{H}_1)\times\hat{n}_1] \\ &=\hat{a}\cdot[\phi_1(\hat{n}_1\times\vec{M}_1)+j\omega\mu_1(\hat{n}_1\times\vec{H}_1)\phi_1]\end{aligned}$$

$$\because\ \vec{a}\cdot(\vec{b}\times\vec{c})=\vec{b}\cdot(\vec{c}\times\vec{a})$$

and based on Eq. (1.19).

Hence,

$$
\begin{aligned}
R.H.S. &= \int_{S_1}\left\{\hat{a}\cdot\left[\nabla'\phi_1\times(\vec{E}_1\times\hat{n}_1)\right]-\hat{a}\cdot\left[\phi_1(\hat{n}_1\times\vec{M}_1)+j\omega\mu_1(\hat{n}_1\times\vec{H}_1)\phi_1\right]\right\}ds' \\
&= \hat{a}\cdot\int_{S_1}\left[-j\omega\mu_1(\hat{n}_1\times\vec{H}_1)\phi_1+(\hat{n}_1\times\vec{E}_1)\times\nabla'\phi_1-(\hat{n}_1\times\vec{M}_1)\phi_1\right]ds'
\end{aligned}
\tag{1.35}
$$

By equating Eqs. (1.34) and (1.35) into Eq. (1.33), we have

$$
\begin{aligned}
&\hat{a}\cdot\int_{V_1}\left[-j\omega\mu_1\vec{J}_1\phi_1-(\nabla'\times\vec{M}_1)\phi_1+(\rho_1/\varepsilon_1)\nabla'\phi_1\right]dv'-\hat{a}\cdot\int_{S_1}(\hat{n}_1\cdot\vec{E}_1)(\nabla'\phi_1)ds' \\
&=\hat{a}\cdot\int_{S_1}\left[-j\omega\mu_1(\hat{n}_1\times\vec{H}_1)\phi_1+(\hat{n}_1\times\vec{E}_1)\times\nabla'\phi_1-(\hat{n}_1\times\vec{M}_1)\phi_1\right]ds'
\end{aligned}
$$

Since $\hat{a}$ is an arbitrary unit vector, it can be dropped from both sides of the above equation. Thus, we have

$$
\begin{aligned}
&\int_{V_1}\left[-j\omega\mu_1\vec{J}_1\phi_1-(\nabla'\times\vec{M}_1)\phi_1+(\rho_1/\varepsilon_1)\nabla'\phi_1\right]dv' \\
&=\int_{S_1}\left[-j\omega\mu_1(\hat{n}_1\times\vec{H}_1)\phi_1+(\hat{n}_1\times\vec{E}_1)\times\nabla'\phi_1+(\hat{n}_1\cdot\vec{E}_1)\nabla'\phi_1-(\hat{n}_1\times\vec{M}_1)\phi_1\right]ds'
\end{aligned}
\tag{1.36}
$$

Equation (1.36) can be simplified further :

$$
\because\quad \nabla'\times(\phi_1\vec{M}_1)=\nabla'\psi_1\times\vec{M}_1+\psi_1\nabla'\times\vec{M}_1
$$

So

$$
\begin{aligned}
\int_{V_1}(\nabla'\times\vec{M}_1)\phi_1 dv' &= \int_{V_1}\left[\nabla'\times(\phi_1\vec{M}_1)\right]dv'-\int_{V_1}\left[\nabla'\phi_1\times\vec{M}_1\right]dv' \\
&= \int_{S_1}\left[\hat{n}_1\times(\phi_1\vec{M}_1)\right]ds'+\int_{V_1}(\vec{M}_1\times\nabla'\phi_1)dv' \\
&= \int_{S_1}\left[(\hat{n}_1\times\vec{M}_1)\phi_1\right]ds'+\int_{V_1}(\vec{M}_1\times\nabla'\phi_1)dv'
\end{aligned}
$$

$$
\because \int_V(\nabla\times\vec{a})dv=\int_S(\hat{n}\times\vec{a})ds
$$

Thus, Eq.(1.36) becomes

$$\int_{V_1}\left[-j\omega\mu_1\vec{J}_1\phi_1-\vec{M}_1\times\nabla'\phi_1+\frac{\rho_1}{\varepsilon_1}\nabla'\phi_1\right]dv'-\int_{S_1}(\hat{n}_1\times\vec{M}_1)\phi_1 ds'$$
$$=\int_{S_1}\left[-j\omega\mu_1(\hat{n}_1\times\vec{H}_1)\phi_1+(\hat{n}_1\times\vec{E}_1)\times\nabla'\phi_1+(\hat{n}_1\cdot\vec{E}_1)\nabla'\phi_1\right]ds'-\int_{S_1}(\hat{n}_1\times\vec{M}_1)\phi_1 ds'$$

Canceling the last terms of the both sides of the above Equation, we have

$$\begin{aligned}&\int_{V_1}\left[-j\omega\mu_1\vec{J}_1\phi_1-\vec{M}_1\times\nabla'\phi_1+\frac{\rho_1}{\varepsilon_1}\nabla'\phi_1\right]dv'\\&=\int_{S_1}\left[-j\omega\mu_1(\hat{n}_1\times\vec{H}_1)\phi_1+(\hat{n}_1\times\vec{E}_1)\times\nabla'\phi_1+(\hat{n}_1\cdot\vec{E}_1)\nabla'\phi_1\right]ds'\end{aligned}\tag{1.37}$$

The surface integral in Eq.(1.37) include three surfaces, or $S_1=S+S_a+S_\infty$.

We will perform surface integral over S_a first.

We will consider,

$$\int_{S_a}\left[-j\omega\mu_1(\hat{n}_1\times\vec{H}_1)\phi_1+(\hat{n}_1\times\vec{E}_1)\times\nabla'\phi_1+(\hat{n}_1\cdot\vec{E}_1)\nabla'\phi_1\right]ds'$$

For the integration over S_a, as shown in Fig.4, the source point (integrating point) $\vec{r}'$ is on the small surface S_a.

$R=|\vec{r}'-\vec{r}_1|$ in the expression of $\phi_1=\dfrac{e^{-j\beta_1 R}}{R}$ is the radius of the small sphere S_a.

$$\phi_1 = \frac{e^{-j\beta_1 R}}{R} = \frac{e^{-j\beta_1|\vec{r}'-\vec{r}_1|}}{|\vec{r}'-\vec{r}_1|}$$

$$\begin{aligned}
\nabla'\phi_1 &= \frac{\partial}{\partial R}\left(\frac{e^{-j\beta_1 R}}{R}\right)\nabla' R \\
&= \left(-\frac{e^{-j\beta_1 R}}{R^2} - j\beta_1\frac{e^{-j\beta_1 R}}{R}\right)\left[\hat{x}\frac{\partial}{\partial x'} + \hat{y}\frac{\partial}{\partial y'} + \hat{z}\frac{\partial}{\partial z'}\right]\left[\sqrt{(x'-x_1)^2+(y'-y_1)^2+(z'-z_1)^2}\right] \\
&= -\left(\frac{1}{R} + j\beta_1\right)\frac{e^{-j\beta_1 R}}{R}\left[\frac{\vec{r}'-\vec{r}_1}{R}\right] \\
&= -\left(\frac{1}{R} + j\beta_1\right)\frac{e^{-j\beta_1 R}}{R}(-\hat{n}_1) \\
&= \left(\frac{1}{R} + j\beta_1\right)\frac{e^{-j\beta_1 R}}{R}\hat{n}_1
\end{aligned}$$

where $\hat{n}_1$ is a unit normal vector on S_a pointing toward to the center of S_a as shown in Fig.4. The first term of the surface integral is

$$\int_{S_a}\left[-j\omega\mu_1(\hat{n}_1\times\vec{H}_1)\phi_1\right]ds' = -j\omega\mu_1(\hat{n}_1\times\vec{H}_1)\lim_{R\to 0}\left[\frac{e^{-j\beta_1 R}}{R}4\pi R^2\right] = 0$$

The second term and the third of the surface integral can be combined as follows :

$$\begin{aligned}
&\int_{S_a}\left[(\hat{n}_1\times\vec{E}_1)\times\nabla'\phi_1 + (\hat{n}_1\cdot\vec{E}_1)\nabla'\phi_1\right]ds' \\
&= \int_{S_a}\left[(\hat{n}_1\times\vec{E}_1)\times\hat{n}_1 + (\hat{n}_1\cdot\vec{E}_1)\hat{n}_1\right]\left(\frac{1}{R} + j\beta_1\right)\frac{e^{-j\beta_1 R}}{R}ds'
\end{aligned}$$

Since,

$$\begin{aligned}
(\hat{n}_1\times\vec{E}_1)\times\hat{n}_1 + (\hat{n}_1\cdot\vec{E}_1)\hat{n}_1 &= -\hat{n}_1\times(\hat{n}_1\times\vec{E}_1) + (\hat{n}_1\cdot\vec{E}_1)\hat{n}_1 \\
&= -(\hat{n}_1\cdot\vec{E}_1)\hat{n}_1 + (\hat{n}_1\cdot\hat{n}_1)\vec{E}_1 + (\hat{n}_1\cdot\vec{E}_1)\hat{n}_1 \\
&= \vec{E}_1
\end{aligned}$$

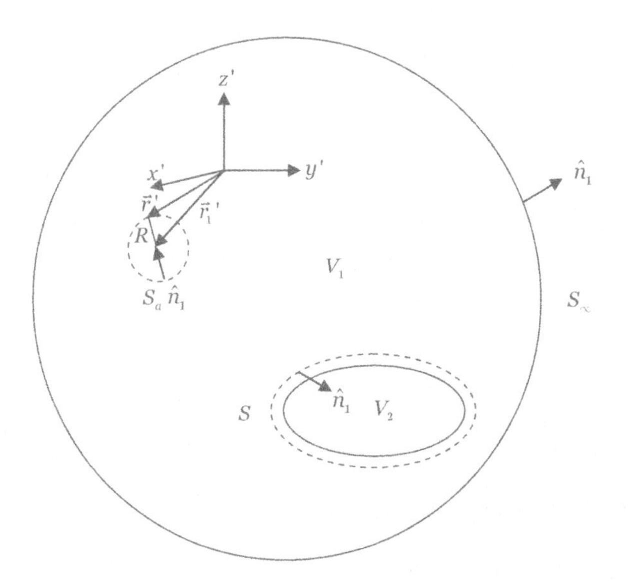

Fig. 4 Surface integration over S_a.

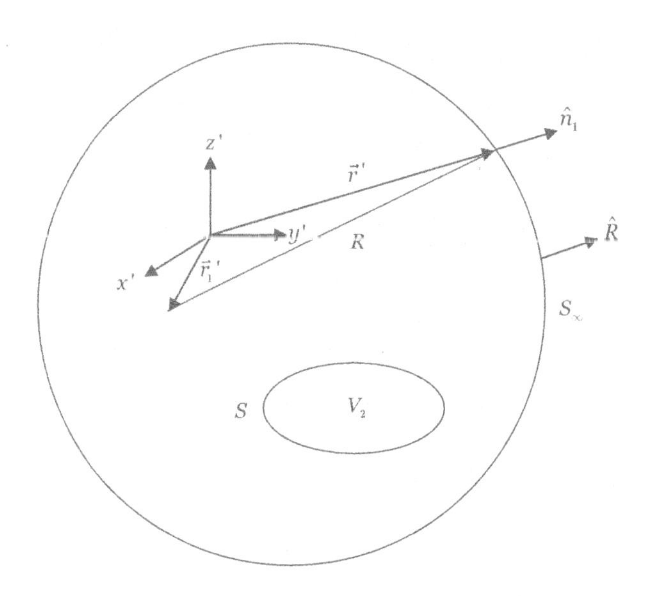

Fig. 5 Surface integration over S_∞.

so

$$\int_{S_a}\left[(\hat{n}_1\times\vec{E}_1)\times\nabla'\phi_1+(\hat{n}_1\cdot\vec{E}_1)\nabla'\phi_1\right]ds'$$

$$=\int_{S_a}\vec{E}_1\left(\frac{1}{R}+j\beta_1\right)\frac{e^{-j\beta_1 R}}{R}ds'$$

$$=\vec{E}_1(\vec{r}_1)\int_{S_a}\left(\frac{1}{R}+j\beta_1\right)\frac{e^{-j\beta_1 R}}{R}ds' \qquad \text{assuming } \vec{E}_1 \text{ is continuous and finite at } \vec{r}_1.$$

$$=\vec{E}_1(\vec{r}_1)\lim_{R\to 0}\left[\left(\frac{1}{R}+j\beta_1\right)\frac{e^{-j\beta_1 R}}{R}4\pi R^2\right]$$

$$=4\pi\vec{E}_1(\vec{r}_1)$$

Therefore,

$$\int_{S_a}\left[-j\omega\mu_1(\hat{n}_1\times\vec{H}_1)\phi_1+(\hat{n}_1\times\vec{E}_1)\times\nabla'\phi_1+(\hat{n}_1\cdot\vec{E}_1)\nabla'\phi_1\right]ds'=4\pi\vec{E}_1(\vec{r}_1) \tag{1.38}$$

Next, let's perform the surface integration of Eq. (1.37) ever S_∞ which is the infinite spherical surface surrounding V_1.

$$\int_{S_\infty}\left[-j\omega\mu_1(\hat{n}_1\times\vec{H}_1)\phi_1+(\hat{n}_1\times\vec{E}_1)\times\nabla'\phi_1+(\hat{n}_1\cdot\vec{E}_1)\nabla'\phi_1\right]ds'$$

$\phi_1=\frac{e^{-j\beta_1 R}}{R}$ and $R=|\vec{r}'-\vec{r}_1|\to\infty$ because $\vec{r}'$ is on the infinite spherical surface S_∞, as shown in Fig. 5.

$$\nabla'\phi_1=\nabla'\left(\frac{e^{-j\beta_1 R}}{R}\right)=\frac{\partial}{\partial R}\left(\frac{e^{-j\beta_1 R}}{R}\right)\nabla' R$$

$$=-\left(\frac{1}{R}+j\beta_1\right)\frac{e^{-j\beta_1 R}}{R}\left[\frac{\vec{r}'-\vec{r}_1}{R}\right]$$

$$=-\left(\frac{1}{R}+j\beta_1\right)\frac{e^{-j\beta_1 R}}{R}(\hat{n}_1) \qquad \hat{n}_1 \text{ is pointing outward from } V_1.$$

$$=-j\beta_1\frac{e^{-j\beta_1 R}}{R}(\hat{R}) \qquad \text{when } R\to\infty \qquad \hat{R} \text{ is he unit radial vector on } S_\infty.$$

Hence,

$$\int_{S_\infty}\left[-j\omega\mu_1(\hat{n}_1\times\vec{H}_1)\phi_1+(\hat{n}_1\times\vec{E}_1)\times\nabla'\phi_1+(\hat{n}_1\cdot\vec{E}_1)\nabla'\phi_1\right]ds'$$
$$=\int_{S_\infty}\left[-j\omega\mu_1(\hat{R}\times\vec{H}_1)-(\hat{R}\times\vec{E}_1)\times\hat{R}j\beta_1-(\hat{R}\cdot\vec{E}_1)\hat{R}j\beta_1\right]\frac{e^{-j\beta_1R}}{R}ds'$$
$$=\int_{S_\infty}\left[-j\omega\mu_1(\hat{R}\times\vec{H}_1)-j\beta_1\vec{E}_1\right]\frac{e^{-j\beta_1R}}{R}ds'$$
$$\because(\hat{R}\times\vec{E}_1)\times\hat{R}+(\hat{R}\cdot\vec{E}_1)\hat{R}=\vec{E}_1$$

Assuming that all the EM sources are confined in a finite volume of V_1 near the origin and that any integrating point $\vec{r}'$ on S_∞ is in the far zone of the EM sources, the $\vec{E}$ and $\vec{H}$ fields on S_∞ should behave as a spherical EM wave and satisfy the following condition.

$$\vec{H}_1=(\hat{R}\times\vec{E}_1)/\sqrt{\frac{\mu_1}{\varepsilon_1}}$$

$$\text{or } \hat{R}\times\vec{H}_1=-\vec{E}_1/\sqrt{\frac{\mu_1}{\varepsilon_1}}$$

and

$$-j\omega\mu_1(\hat{R}\times\vec{H}_1)-j\beta_1\vec{E}_1=j\omega\mu_1\vec{E}_1/\sqrt{\frac{\mu_1}{\varepsilon_1}}-j\beta_1\vec{E}_1=0$$

hence

$$\int_{S_\infty}\left[-j\omega\mu_1(\hat{n}_1\times\vec{H}_1)\phi_1+(\hat{n}_1\times\vec{E}_1)\times\nabla'\phi_1+(\hat{n}_1\cdot\vec{E}_1)\nabla'\phi_1\right]ds'=0 \tag{1.39}$$

With Eq. (1.38) and Eq. (1.39), the surface integral of Eq. (1.37) becomes

$$\begin{aligned}&\int_{S_1=S+S_a+S_\infty}\left[-j\omega\mu_1(\hat{n}_1\times\vec{H}_1)\phi_1+(\hat{n}_1\times\vec{E}_1)\times\nabla'\phi_1+(\hat{n}_1\cdot\vec{E}_1)\nabla'\phi_1\right]ds'\\&=4\pi\vec{E}_1(\vec{r}_1)+\int_S\left[-j\omega\mu_1(\hat{n}_1\times\vec{H}_1)\phi_1+(\hat{n}_1\times\vec{E}_1)\times\nabla'\phi_1+(\hat{n}_1\cdot\vec{E}_1)\nabla'\phi_1\right]ds'\end{aligned} \tag{1.40}$$

Substituting Eq. (1.40), in Eq. (1.37), we have

$$\begin{aligned}\vec{E}_1(\vec{r}_1) &= \frac{1}{4\pi}\int_{V_1}\left[-j\omega\mu_1\vec{J}_1\phi_1 - \vec{M}_1\times\nabla'\phi_1 + \frac{\rho_1}{\varepsilon_1}\nabla'\phi_1\right]dv' \\ &-\frac{1}{4\pi}\int_S\left[-j\omega\mu_1(\hat{n}_1\times\vec{H}_1)\phi_1 + (\hat{n}_1\times\vec{E}_1)\times\nabla'\phi_1 + (\hat{n}_1\cdot\vec{E}_1)\nabla'\phi_1\right]ds'\end{aligned} \tag{1.41}$$

At this point, we can define the equivalent electric and magnetic surface current as

$$\vec{J}_s = -\vec{n}_1\times\vec{H}_1 = \hat{n}\times\vec{H}_1 \tag{1.42}$$

$$\vec{M}_s = \vec{n}_1\times\vec{E}_1 = -\hat{n}\times\vec{E}_1 \tag{1.43}$$

where $\hat{n}$ is the unit vector pointing outward from region 2 on S, and $\hat{n}_1$ is the outgoing unit vector of region 1 on S. We can also define the equivalent electric and surface charge as follows :

From Eq. (1.19), $\hat{n}_1\cdot(\nabla\times\vec{H}_1) = \hat{n}_1\cdot\vec{J}_1 + j\omega\varepsilon_1\hat{n}_1\cdot\vec{E}_1$

If no source current $\vec{J}_1$ is present at S, we have

$$\begin{aligned}\hat{n}_1\cdot\vec{E}_1 &= \frac{-j}{\omega\varepsilon_1}n_1\cdot(\nabla\times\vec{H}_1) = \frac{j}{\omega\varepsilon_1}\nabla\cdot(\hat{n}_1\times\vec{H}_1) \\ &= \frac{-j}{\omega\varepsilon_1}\nabla\cdot\vec{J}_S = \frac{-1}{\varepsilon_1}\rho_S\end{aligned} \tag{1.44}$$

where ρ_s is the equivalent electric surface charge associated with $\vec{J}_s$ by the continuity equation of $\nabla\cdot\vec{J}_s + j\omega\rho_s = 0$.

Substituting Eq. (1.42) to Eq. (1.44) into Eq. (1.41) leads to

$$\begin{aligned}\vec{E}_1(\vec{r}_1) &= \frac{1}{4\pi}\int_{V_1}\left[-j\omega\mu_1\vec{J}_1\phi_1 - \vec{M}_1\times\nabla'\phi_1 + \frac{\rho_1}{\varepsilon_1}\nabla'\phi_1\right]dv' \\ &+\frac{1}{4\pi}\int_S\left[-j\omega\mu_1\vec{J}_S\phi_1 - \vec{M}_S\times\nabla'\phi_1 + \frac{\rho_S}{\varepsilon_1}\nabla'\phi_1\right]ds'\end{aligned} \tag{1.45}$$

Examine the three terms in the integrand of the volume integral and compare with the three corresponding terms in the surface integral, obviously they represent the contributions due to the electric current, the magnetic current and the electric charge, respectively. However, those surface currents and charge are equivalent sources derived from the component of the $\vec{E}_1$ and $\vec{H}_1$ on the surface S of the region 2. The physical meaning of Eq. (1.45) is as follows :

The electric field at an interior point $\vec{r}'$ within V_1, $\vec{E}_1(\vec{r}_1)$, is maintained by the given source currents $(\vec{J}_1, \vec{M}_1)$ in V_1 and equivalent surface currents $(\vec{J}_s, \vec{M}_s)$ on the surface S while the medium of region 2 is replaced by that of region 1 and the source currents $(\vec{J}_2, \vec{M}_2)$ in V_2 are removed. This is because the parameters (ε_2, μ_2) and $(\vec{J}_2, \vec{M}_2)$ do not appear in Eq. (1.45) and the unbound Green's function ϕ_1 appears in both the volume and surface integrals in Eq. (1.45). From the appearance of Eq. (1.45), $\vec{E}_1(\vec{r}_1)$ is maintained by the source currents $(\vec{J}_1, \vec{M}_1)$ and the equivalent surface currents $(\vec{J}_s, \vec{M}_s)$ located in the unbounded homogeneous region with electric parameters of (ε_1, μ_1).

It is to be reminded that the equivalent surface currents $(\vec{J}_s, \vec{M}_s)$ on the surface S are still unknown because $\vec{E}_1$ and $\vec{H}_1$ on S can be determined from the matching of the boundary conditions between regions 1 and 2, as will be demonstrated later.

Next, let us consider the case when the field point $\vec{r}$ is on the surface S, such as $\vec{r}_2$ in Fig. 2. For this case, we need to exclude the singularity point r_2 from V_1 with a hemisphere which has a hemispherical surface S_a as shown in Fig. 2 before we can use Eq. (1.28). With this S_a, the surface integral over S_a in Eq. (1.37) becomes

$$\int_{S_a} [............] ds' = 2\pi \vec{E}_1(\vec{r}_2) \tag{1.46}$$

Therefore Eq. (1.37) can be rearranged to give $\vec{E}_1(\vec{r}_2)$ as

$$\begin{aligned}\vec{E}_1(\vec{r}_2)=&\frac{1}{2\pi}\int_{V_1}\left[-j\omega\mu_1\vec{J}_1\phi_1-\vec{M}_1\times\nabla'\phi_1+\frac{\rho_1}{\varepsilon_1}\nabla'\phi_1\right]dv'\\&+\frac{1}{2\pi}\int_{S}\left[-j\omega\mu_1\vec{J}_S\phi_1-\vec{M}_S\times\nabla'\phi_1+\frac{\rho_S}{\varepsilon_1}\nabla'\phi_1\right]ds'\end{aligned}\tag{1.47}$$

Comparing Eq. (1.47) with Eq. (1.45), there is a factor of 2 between them. The surface integral in Eq. (1.47) is a principal value integral which excludes the contribution from the singularity point $\vec{r}_2$. The Green's function ϕ_1 in Eq. (1.47) is

$$\phi_1=\exp\left(-j\beta_1|\vec{r}'-\vec{r}_2|\right)/|\vec{r}'-\vec{r}_2|$$

We will now proceed to derive the magnetic field, $\vec{H}_1(\vec{r}_1)$, in the region 1. If we let,

$$\vec{P}=\vec{H}_1(\vec{r}')\tag{1.48}$$

and

$$\vec{Q}=\hat{a}\phi_1(\vec{r}')$$

same as Eq. (1.30), and proceed to go through the same lengthy derivation as we have performed for $\vec{E}_1(\vec{r}_1)$, we will obtain the following results :

$$\begin{aligned}\vec{H}_1(\vec{r}_1)=&\frac{1}{4\pi}\int_{V_1}\left[-j\omega\varepsilon_1\vec{M}_1\phi_1+\vec{J}_1\times\nabla'\phi_1+\frac{\rho_{m1}}{\mu_1}\nabla'\phi_1\right]dv'\\&+\frac{1}{4\pi}\int_{S}\left[-j\omega\varepsilon_1\vec{M}_S\phi_1+\vec{J}_S\times\nabla'\phi_1+\frac{\rho_{ms}}{\mu_1}\nabla'\phi_1\right]ds'\end{aligned}\tag{1.49}$$

for $\vec{r}=\vec{r}_1$ (interior point within V_1)

and

$$\vec{H}_1(\vec{r}_2)=\frac{1}{2\pi}\int_{V_1}\left[-j\omega\varepsilon_1\vec{M}_1\phi_1+\vec{J}_1\times\nabla'\phi_1+\frac{\rho_{m1}}{\mu_1}\nabla'\phi_1\right]dv'$$
$$+\frac{1}{2\pi}\int_{S}\left[-j\omega\varepsilon_1\vec{M}_s\phi_1+\vec{J}_s\times\nabla'\phi_1+\frac{\rho_{ms}}{\mu_1}\nabla'\phi_1\right]ds' \tag{1.50}$$

for $\vec{r}=\vec{r}_2$ (surface point on S)

where

$$\rho_{m1}=\frac{j}{\omega}\nabla\cdot\vec{M}_1 \text{ and } \rho_{ms}=\frac{j}{\omega}\nabla\cdot\vec{M}_s \tag{1.51}$$

$\vec{M}_s$ has been defined in Eq. (1.43)

Next, we can repeat a similar derivation for region 2, choosing

$$\vec{P}(\vec{r}')=\vec{E}_2(\vec{r}') \tag{1.52}$$

$$\vec{Q}(\vec{r}')=\hat{a}\phi_2(\vec{r}',\vec{r})=\hat{a}\exp\left[\left(-j\beta_2\,|\vec{r}'-\vec{r}\,|\right)/\,|\vec{r}'-\vec{r}\,|\right] \tag{1.53}$$

where $\beta_2=\omega\sqrt{\mu_2\varepsilon_2}$, and substituting $\vec{P}$ and $\vec{Q}$ into Eq. (1.28), we will have

$$\int_{V_2}\left[-j\omega\mu_2\vec{J}_2\phi_2-\vec{M}_2\times\nabla'\phi_2+\frac{\rho_2}{\varepsilon_2}\nabla'\phi_2\right]dv'$$
$$=\int_{S_2}\left[-j\omega\mu_2(\hat{n}_2\times\vec{H}_2)\phi_2+(\hat{n}_2\times\vec{E}_2)\times\nabla'\phi_2+(\hat{n}_2\cdot\vec{E}_2)\nabla'\phi_2\right]ds' \tag{1.54}$$

The total boundary surface S_2 for the finite volume V_2 is

$$S_2 = S + S_a$$

where S_a is the surface of a small sphere (or hemisphere) for excluding the singularity point $\vec{r}$ in ϕ_2. It is noted that the infinite spherical surface S_∞ is not needed because V_2 is a finite volume. Also $\hat{n}_2$ is the normal unit vector pointing outward from V_2 as shown in Fig. 2.

Following the same manipulation used for the case of region 1, we can obtain $\vec{E}_2(\vec{r})$ at an interior point within V_2 as

$$\begin{aligned}\vec{E}_2(\vec{r}) = &\frac{1}{4\pi}\int_{V_2}\left[-j\omega\mu_2\vec{J}_2\phi_2 - \vec{M}_2\times\nabla'\phi_2 + \frac{\rho_2}{\varepsilon_2}\nabla'\phi_2\right]dv' \\ &-\frac{1}{4\pi}\int_S\left[-j\omega\mu_2(\hat{n}_2\times\vec{H}_2)\phi_2 + (\hat{n}_2\times\vec{E}_2)\times\nabla'\phi_2 + (\hat{n}_2\cdot\vec{E}_2)\nabla'\phi_2\right]ds'\end{aligned} \tag{1.55}$$

similar to Eq. (1.41) for $\vec{E}_1(\vec{r}_1)$.

We will use the definition of the equivalent surface currents $(\vec{J}_S, \vec{M}_S)$ given in Eq. (1.42) and Eq. (1.43) :

$$\vec{J}_S = -\hat{n}_1\times\vec{H}_1 = \hat{n}\times\vec{H}_1 = \hat{n}\times\vec{H}_2 = \hat{n}_2\times\vec{H}_2$$
$$\vec{M}_S = \hat{n}_1\times\vec{E}_1 = -\hat{n}\times\vec{E}_1 = -\hat{n}\times\vec{E}_2 = -\hat{n}_2\times\vec{E}_2$$

Because the tangential components of $\vec{E}$ and $\vec{H}$ fields are continuous across S :

$$\hat{n}\times\vec{E}_1 = \hat{n}\times\vec{E}_2$$
$$\hat{n}\times\vec{H}_1 = \hat{n}\times\vec{H}_2$$

and $$\hat{n}_1 = -\hat{n}\ ,\quad \hat{n}_2 = \hat{n}$$

we can now rewrite Eq. (1.55) as

$$\vec{E}_2(\vec{r}) = \frac{1}{4\pi}\int_{V_2}\left[-j\omega\mu_2\vec{J}_2\phi_2 - \vec{M}_2\times\nabla'\phi_2 + \frac{\rho_2}{\varepsilon_2}\nabla'\phi_2\right]dv'$$
$$+\frac{1}{4\pi}\int_{S}\left[-j\omega\mu_2(-\vec{J}_S)\phi_2 - (-\vec{M}_S)\times\nabla'\phi_2 + \frac{-\rho_S}{\varepsilon_2}\nabla'\phi_2\right]ds' \qquad (1.56)$$

($\vec{r}$ is an interior point within V_2)

Notice that equivalent surface current which can maintain the correct $\vec{E}$ field inside V_2 are $(-\vec{J}_S, -\vec{M}_S)$ which flow in opposite directions on S compared with case of region 1. Eq. (1.56) implies that when the source currents $(\vec{J}_1, \vec{M}_1)$ in V_1 are removed and medium of region 1 is replaced by that of region 2 (to make the whole space homogeneous), the correct value of the electric field at an interior point $\vec{r}$ inside V_2 can be calculated from the source currents $(\vec{J}_2, \vec{M}_2)$ in V_2 and the negative equivalent surface currents $(-\vec{J}_S, -\vec{M}_S)$ on S.

Similarly, the electric field at a field point on S can be expressed as

$$\vec{E}_2(\vec{r}) = \frac{1}{2\pi}\int_{V_2}\left[-j\omega\mu_2\vec{J}_2\phi_2 - \vec{M}_2\times\nabla'\phi_2 + \frac{\rho_2}{\varepsilon_2}\nabla'\phi_2\right]dv'$$
$$+\frac{1}{2\pi}\int_{S}\left[-j\omega\mu_2(-\vec{J}_S)\phi_2 - (-\vec{M}_S)\times\nabla'\phi_2 + \frac{-\rho_S}{\varepsilon_S}\nabla'\phi_2\right]ds' \qquad (1.57)$$

($\vec{r}$ is on the surface S)

The results for the $\vec{H}$ field in region 2, $\vec{H}_2(\vec{r})$, are similar to those given in Eq. (1.49) and Eq. (1.50).

As stated earlier, the EM field in regions 1 and 2, $\left[\vec{E}_1(\vec{r}),\vec{H}_1(\vec{r})\right]$ and $\left[\vec{E}_2(\vec{r}),\vec{H}_2(\vec{r})\right]$, can be explicitly calculated once the equivalent surface currents, $\vec{J}_S$ and $\vec{M}_S$, on the boundary surface S are determined. They can be determined by matching the tangential components of $\vec{E}$ and $\vec{H}$ fields on the surface S:

$$\begin{cases} \hat{n}\times\vec{E}_1(\vec{r})=\hat{n}\times\vec{E}_2(\vec{r}) \\ \hat{n}\times\vec{H}_1(\vec{r})=\hat{n}\times\vec{H}_2(\vec{r}) \end{cases}$$

when $\vec{r}$ is on S

From Eq. (1.47),

$$\begin{aligned} \hat{n}\times\vec{E}_1(\vec{r}) &= \hat{n}\times\vec{E}_1(\vec{r}_2) \\ &= \hat{n}\times\frac{1}{2\pi}\int_{V_1}\left[-j\omega\mu_1\vec{J}_1\phi_1-\vec{M}_1\times\nabla'\phi_1+\frac{\rho_1}{\varepsilon_1}\nabla'\phi_1\right]dv' \\ &\quad +\hat{n}\times\frac{1}{2\pi}\int_{S}\left[-j\omega\mu_1\vec{J}_S\phi_1-\vec{M}_S\times\nabla'\phi_1+\frac{\rho_S}{\varepsilon_1}\nabla'\phi_1\right]ds' \end{aligned} \tag{1.58}$$

where $\vec{r}$ is on S

From Eq. (1.57), we can write,

$$\begin{aligned} \hat{n}\times\vec{E}_2(\vec{r}) &= \hat{n}\times\frac{1}{2\pi}\int_{V_2}\left[-j\omega\mu_2\vec{J}_2\phi_2-\vec{M}_2\times\nabla'\phi_2+\frac{\rho_2}{\varepsilon_2}\nabla'\phi_2\right]dv' \\ &\quad +\hat{n}\times\frac{1}{2\pi}\int_{S}\left[-j\omega\mu_2(-\vec{J}_S)\phi_2-(-\vec{M}_S)\times\nabla'\phi_2-\frac{\rho_S}{\varepsilon_2}\nabla'\phi_2\right]ds' \end{aligned} \tag{1.59}$$

where $\vec{r}$ is on S

From Eq. (1.50),

$$\begin{aligned}\hat{n}\times\vec{H}_1(\vec{r}) &= \hat{n}\times\vec{H}_1(\vec{r}_2)\\ &= \hat{n}\times\frac{1}{2\pi}\int_{V_1}\left[-j\omega\varepsilon_1\vec{M}_1\phi_1+\vec{J}_1\times\nabla'\phi_1+\frac{\rho_{m1}}{\mu_1}\nabla'\phi_1\right]dv'\\ &\quad+\hat{n}\times\frac{1}{2\pi}\int_{S}\left[-j\omega\varepsilon_1\vec{M}_s\phi_1+\vec{J}_s\times\nabla'\phi_1+\frac{\rho_{ms}}{\mu_1}\nabla'\phi_1\right]ds'\end{aligned} \tag{1.60}$$

where $\vec{r}$ is on S.

Similarly we can write,

$$\begin{aligned}\hat{n}\times\vec{H}_2(\vec{r}) &= \hat{n}\times\frac{1}{2\pi}\int_{V_2}\left[-j\omega\varepsilon_2\vec{M}_2\phi_2+\vec{J}_2\times\nabla'\phi_2+\frac{\rho_{m2}}{\mu_2}\nabla'\phi_2\right]dv'\\ &\quad+\hat{n}\times\frac{1}{2\pi}\int_{S}\left[-j\omega\varepsilon_2(-\vec{M}_s)\phi_2+(-\vec{J}_s)\times\nabla'\phi_2+\frac{-\rho_{ms}}{\mu_2}\nabla'\phi_2\right]ds'\end{aligned} \tag{1.61}$$

where $\vec{r}$ is on S.

By matching $\hat{n}\times\vec{E}_1(\vec{r})=\hat{n}\times\vec{E}_2(\vec{r})$ on S, we have

$$\begin{aligned}&\hat{n}\times\frac{1}{2\pi}\int_{S}\left[-j\omega(\mu_1\phi_1+\mu_2\phi_2)\vec{J}_S-\vec{M}_S\times(\nabla'\phi_1+\nabla'\phi_2)+\rho_S\left(\frac{\nabla'\phi_1}{\varepsilon_1}+\frac{\nabla'\phi_2}{\varepsilon_2}\right)\right]ds'\\ &=-\hat{n}\times\frac{1}{2\pi}\int_{V_1}\left[-j\omega\mu_1\vec{J}_1\phi_1-\vec{M}_1\times\nabla'\phi_1+\frac{\rho_1}{\varepsilon_1}\nabla'\phi_1\right]dv'\\ &\quad+\hat{n}\times\frac{1}{2\pi}\int_{V_2}\left[-j\omega\mu_2\vec{J}_2\phi_2-\vec{M}_2\times\nabla'\phi_2+\frac{\rho_2}{\varepsilon_2}\nabla'\phi_2\right]dv'\end{aligned} \tag{1.62}$$

The RHS of Eq. (1.62) can be evaluated easily if $(\vec{J}_1, \vec{M}_1)$ and $(\vec{J}_2, \vec{M}_2)$ are specified, noting that ρ_1 and ρ_2 can be derived from $\vec{J}_1$ and $\vec{J}_2$ based on the continuity equation. The LHS of Eq. (1.62) contains two unknowns, $\vec{J}_S$ and $\vec{M}_S$, which are to be determined. It is noted that ρ_s is not a new unknown because it can be derived from $\vec{J}_S$ using the continuity equation. Since we have two unknowns we need one more equation for $\vec{J}_S$ and $\vec{M}_S$ to completely determined them. The second equation can be obtained by matching the tangential components of $\vec{H}$ field on the boundary surface S. By matching $\hat{n} \times \vec{H}_1(\vec{r}) = \hat{n} \times \vec{H}_2(\vec{r})$ on S, we have

$$\begin{aligned}
&\hat{n} \times \frac{1}{2\pi} \int_S \left[-j\omega(\varepsilon_1 \phi_1 + \varepsilon_2 \phi_2)\vec{M}_S + \vec{J}_S \times (\nabla' \phi_1 + \nabla' \phi_2) + \rho_{mS} \left(\frac{\nabla' \phi_1}{\mu_1} + \frac{\nabla' \phi_2}{\mu_2} \right) \right] ds' \\
&= -\hat{n} \times \frac{1}{2\pi} \int_{V_1} \left[-j\omega\varepsilon_1 \vec{M}_1 \phi_1 + \vec{J}_1 \times \nabla' \phi_1 + \frac{\rho_{m1}}{\mu_1} \nabla' \phi_1 \right] dv' \\
&+ \hat{n} \times \frac{1}{2\pi} \int_{V_2} \left[-j\omega\varepsilon_2 \vec{M}_2 \phi_2 + \vec{J}_2 \times \nabla' \phi_2 + \frac{\rho_{m2}}{\mu_2} \nabla' \phi_2 \right] dv'
\end{aligned} \tag{1.63}$$

Now, $\vec{J}_S$ and $\vec{M}_S$ can be determined numerically from Eqs. (1.62) and (1.63). After that $\left[\vec{E}_1(\vec{r}), \vec{H}_1(\vec{r})\right]$ and $\left[\vec{E}_2(\vec{r}), \vec{H}_2(\vec{r})\right]$ can be calculated from Eqs. (1.45), (1.47), (1.49), (1.50), (1.56) and (1.57).

The solutions for $\vec{E}_1$ and $\vec{H}_1$ fields derived in this section may be the most general solutions of general Maxwell's equations applied in heterogeneous multiple media.

1.c Mathematical Derivation of Equivalence Principle

The equivalence principle in electromagnetics has been well known for a long time, having been presented by Harrington [1] in a descriptive manner in his book. Recently, this principle has found many applications in problems involving the interaction of EM fields with material bodies. In these applications, accurate mathematical formulation of the equivalence principle are needed. In this section a mathematical formulation of the equivalence principle will be presented as the extension of the results derived in the preceding section.

Let's consider the same geometry as depicted in Fig.2 of the preceding section. This geometry consists of region 2 with complex permittivity and permeability (ε_2, μ_2), the volume V_2, the boundary surface S, and the electric and magnetic source currents $(\vec{J}_2, \vec{M}_2)$ within V_2. Region 2 is surrounded by region 1 of infinite volume V_1 that has electric parameters of (ε_1, μ_1) and source currents of $(\vec{J}_1, \vec{M}_1)$ within V_1.

Using Maxwell's equations and vector Green's theorem and assuming

$$\vec{P} = \vec{E}_1(\vec{r}\,') \text{ and } \vec{Q} = \hat{a}\phi_1(\vec{r}\,', \vec{r})$$

We have obtained the electric field at an interior point $\vec{r} = \vec{r}_1$ within V_1, $\vec{E}_1(\vec{r}_1)$, as maintained by $(\vec{J}_1, \vec{M}_1)$ and the equivalent surface currents $(\vec{J}_S, \vec{M}_S)$ as

$$\begin{aligned}\vec{E}_1(\vec{r}_1) = &\frac{1}{4\pi}\int_{V_1}[-j\omega\mu_1\vec{J}_1\phi_1 - \vec{M}_1 \times \nabla'\phi_1 + \frac{\rho_1}{\varepsilon_1}\nabla'\phi_1]dv' \\ &+ \frac{1}{4\pi}\int_S[-j\omega\mu_1\vec{J}_S\phi_1 - \vec{M}_S \times \nabla'\phi_1 + \frac{\rho_S}{\varepsilon_1}\nabla'\phi_1]ds'\end{aligned} \tag{1.45}$$

Now, we want to follow the same derivation leading to Eq.(1.45), but changing

$$\vec{Q} = \hat{a}\phi_1(\vec{r}\,', \vec{r}_3) = \hat{a}\exp(j\beta|\vec{r}\,' - \vec{r}_3|)/|\vec{r}\,' - \vec{r}_3|$$

or changing the field point $\vec{r} = \vec{r}_3$ to be located in region 2 instead of region 1, as shown in Fig.2. Omitting the lengthy derivation in the preceding section and use Eq.(1.37)

$$\begin{aligned}&\int_{V_1}\left[-j\omega\mu_1\vec{J}_1\phi_1 - \vec{M}_1\times\nabla'\phi_1 + \frac{\rho_1}{\varepsilon_1}\nabla'\phi_1\right]dv' \\ &= \int_{S_1}\left[-j\omega\mu_1(\hat{n}_1\times\vec{H}_1)\phi_1 + (\hat{n}_1\times\vec{E}_1)\times\nabla'\phi_1 + (\hat{n}_1\cdot\vec{E}_1)\nabla'\phi_1\right]ds'\end{aligned} \tag{1.37}$$

Originally $S_1 = S + S_a + S_\infty$, but in this case, the field point or the singularity point $\vec{r}_3$ in $\vec{Q}$ function is outside V_1. Thus, $\vec{Q}$ function is continuous everywhere within V_1 and there is no need to isolate $\vec{r}_3$ with a small sphere S_a. Therefore, the surrounding boundary surface for V_1 is $S_1 = S + S_\infty$.

Since the surface integral over the infinite surface S_∞ can be shown to be zero due to the radiation condition, Eq.(1.37) is reduced to

$$\begin{aligned}&\int_{V_1}[-j\omega\mu_1\vec{J}_1\phi_1 - \vec{M}_1\times\nabla'\phi_1 + \frac{\rho_1}{\varepsilon_1}\nabla'\phi_1]dv' \\ &= \int_{S}[-j\omega\mu_1(\hat{n}_1\times\vec{H}_1)\phi_1 + (\hat{n}_1\times\vec{E}_1)\times\nabla'\phi_1 + (\hat{n}_1\cdot\vec{E}_1)\nabla'\phi_1]ds' \\ &= -\int_{S}[-j\omega\mu_1\vec{J}_S\phi_1 - \vec{M}_S\times\nabla'\phi_1 + \frac{\rho_S}{\varepsilon_1}\nabla'\phi_1]ds'\end{aligned} \tag{1.64}$$

for $\vec{r} = \vec{r}_3$, and based on the definition of the equivalent surface current and charge given in Eqs. (1.42) to (1.44).

Now, if we try to express the electric field at $\vec{r}_3$ maintained by the given source currents $(\vec{J}_1, \vec{M}_1)$ in V_1 and the equivalent surface currents $(\vec{J}_S, \vec{M}_S)$ on S while replacing the medium in region 2 with that of region 1 and removing the source currents $(\vec{J}_2, \vec{M}_2)$ in V_2, we should have an expression for $\vec{E}_2(\vec{r}_3)$ with the following form:

$$\begin{aligned}\vec{E}_2(\vec{r}_3) = & \frac{1}{4\pi}\int_{V_1}[-j\omega\mu_1\vec{J}_1\phi_1 - \vec{M}_1\times\nabla'\phi_1 + \frac{\rho_1}{\varepsilon_1}\nabla'\phi_1]dv' \\ & + \frac{1}{4\pi}\int_{S}[-j\omega\mu_1\vec{J}_S\phi_1 - \vec{M}_S\times\nabla'\phi_1 + \frac{\rho_S}{\varepsilon_1}\nabla'\phi_1]ds'\end{aligned} \tag{1.65}$$

with $\vec{r} = \vec{r}_3$.

Combining Eqs. (1.64) and (1.65) we have

$$\vec{E}_2(\vec{r}_3) = 0 \tag{1.66}$$

This is an interesting result. It means that if the source currents $(\vec{J}_2, \vec{M}_2)$ in V_2 are removed and the medium of region 2 is replaced by that of region 1 (to make the whole space homogeneous), then the source currents $(\vec{J}_1, \vec{M}_1)$ in V_1 and the equivalent surface currents $(\vec{J}_S, \vec{M}_S)$ on S will maintain a zero electric field at any point within region 2.

We have derived the $\vec{H}$ field maintained by the given source currents $(\vec{J}_1, \vec{M}_1)$ in V_1 and the equivalent surface current $(\vec{J}_S, \vec{M}_S)$ on S at an interior point $\vec{r}_1$ as

$$\begin{aligned}\vec{H}_1(\vec{r}_1) = & \frac{1}{4\pi}\int_{V_1}\left[-j\omega\varepsilon_1\vec{M}_1\phi_1 + \vec{J}_1\times\nabla'\phi_1 + \frac{\rho_{m1}}{\mu_1}\nabla'\phi_1\right]dv' \\ & + \frac{1}{4\pi}\int_{S}\left[-j\omega\varepsilon_1\vec{M}_S\phi_1 + \vec{J}_S\times\nabla'\phi_1 + \frac{\rho_{ms}}{\mu_1}\nabla'\phi_1\right]ds'\end{aligned} \tag{1.49}$$

for $\vec{r} = \vec{r}_1$ (interior point within V_1)

If we try to find the $\vec{H}$ field at a field point $\vec{r}_3$ located within V_2 which is maintained by $(\vec{J}_1, \vec{M}_1)$ in V_1 and $(\vec{J}_S, \vec{M}_S)$ on S, we should arrive at the result of

$$\vec{H}_2(\vec{r}_3) = 0 \tag{1.67}$$

for $\vec{r} = \vec{r}_3$

This is the similar result as for the $\vec{E}$ field as shown in Eq. (1.66). We have so far derived a half of the equivalent principle.

Next, we will move to region 2. We have obtained the electric field at an interior point within V_2, $\vec{E}_2(\vec{r})$,as

$$\begin{aligned}\vec{E}_2(\vec{r}) &= \frac{1}{4\pi}\int_{V_2}[-j\omega\mu_2\vec{J}_2\phi_2 - \vec{M}_2 \times \nabla'\phi_2 + \frac{\rho_2}{\varepsilon_2}\nabla'\phi_2]dv' \\ &+ \frac{1}{4\pi}\int_S[-j\omega\mu_2(-\vec{J}_S)\phi_2 - (-\vec{M}_S)\times\nabla'\phi_2 + \frac{-\rho_S}{\varepsilon_2}\nabla'\phi_2]ds'\end{aligned} \tag{1.56}$$

$\vec{r}$ is an interior point within V_2, and $\phi_2 = exp(-j\beta_2|\vec{r}' - \vec{r}|)/|\vec{r}' - \vec{r}|$

If we choose the field point to be outside of V_2 and located in V_1, Eq. (1.54) in the preceding section will give

$$\begin{aligned}&\int_{V_2}[-j\omega\mu_2\vec{J}_2\phi_2 - \vec{M}_2 \times \nabla'\phi_2 + \frac{\rho_2}{\varepsilon_2}\nabla'\phi_2]dv' \\ &= \int_{S_2=S}[-j\omega\mu_2(\hat{n}_2 \times \vec{H}_2)\phi_2 + (\hat{n}_2 \times \vec{E}_2)\times\nabla'\phi_2 + (\hat{n}_2 \cdot \vec{E}_2)\nabla'\phi_2]ds' \\ &= -\int_S[-j\omega\mu_2(-\vec{J}_S)\phi_2 - (-\vec{M}_S)\times\nabla'\phi_2 + \frac{-\rho_S}{\varepsilon_2}\nabla'\phi_2]ds'\end{aligned} \tag{1.68}$$

based on the definitions of the equivalent surface currents and changes. Also it is due to the fact that the field point or the singularity point is located outside of or inside V_1, thus, ϕ_2 is continuous everywhere within V_2 and we don't need to create S_a to isolate the field point. So that $S_2 = S$ only.

Now, if we attempt to express the electric field at a field point $\vec{r}_3$ located in V_1 and which is maintained by the source current $(\vec{J}_2, \vec{M}_2)$ in V_2 and the negative equivalent surface currents $(-\vec{J}_S, -\vec{M}_S)$ on S, while removing the source currents $(\vec{J}_1, \vec{M}_1)$ and replacing the medium of region 1 with that of region 2 (to make the whole space homogeneous), we should have the following expression :

$$\begin{aligned}\vec{E}_1(\vec{r}_3) = & \frac{1}{4\pi}\int_{V_2}[-j\omega\mu_2\vec{J}_2\phi_2 - \vec{M}_2\times\nabla'\phi_2 + \frac{\rho_2}{\varepsilon_2}\nabla'\phi_2]dv' \\ & + \frac{1}{4\pi}\int_{S}[-j\omega\mu_2(-\vec{J}_S)\phi_2 - (-\vec{M}_S)\times\nabla'\phi_2 + \frac{-\rho_S}{\varepsilon_2}\nabla'\phi_2]ds'\end{aligned} \quad (1.69)$$

$\vec{r}_3$ is outside of V_2 and in V_1.

The substitution of Eq.(1.68) in Eq.(1.69) will lead to

$$\vec{E}_1(\vec{r}_3) = 0 \quad (1.70)$$

where $\vec{r}_3$ is outside of V_2 and in V_1.

This confirms the result that the source currents $(\vec{J}_2, \vec{M}_2)$ in V_2 and the negative equivalent surface current $(-\vec{J}_S, -\vec{M}_S)$ on S will maintains a zero electric field at any point in V_1 after removing the source current $(\vec{J}_1, \vec{M}_1)$ in V_1 and fill the whole space with the medium of region 2.

We can easily derive similar results for the $\vec{H}$ field maintained by the source currents $(\vec{J}_2, \vec{M}_2)$ in V_2 and the negative equivalent surface currents $(-\vec{J}_S, -\vec{M}_S)$ on S at an interior point within V_2, and a zero $\vec{H}$ field inside V_1.

These results represent the essence of the equivalence principle. This principle is graphically explained in Figs.6 and 7. The material covered in this section has been presented by Chen [2] in his paper in 1989.

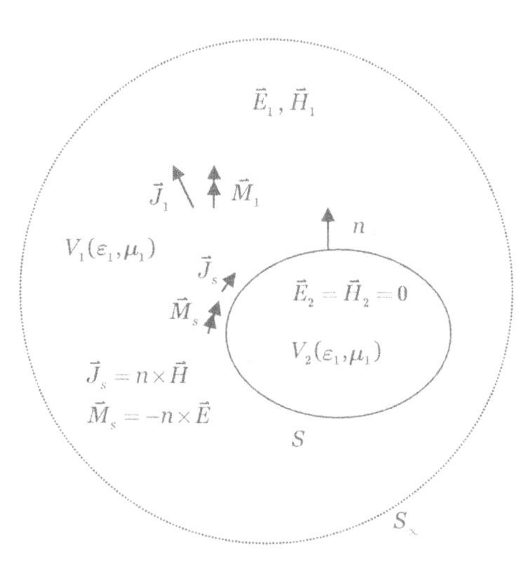

Fig. 6 When the source currents $(\vec{J}_2, \vec{M}_2)$ in V_2 are removed and the medium of region 2 is replaced with that of region 1, the source currents $(\vec{J}_1, \vec{M}_1)$ in V_1 and the equivalent surface currents $(\vec{J}_s, \vec{M}_s)$ on S will maintain the correct EM field $(\vec{E}_1, \vec{H}_1)$ in V_1 and zero EM field ($\vec{E}_2 = \vec{H}_2 = 0$) in V_2.

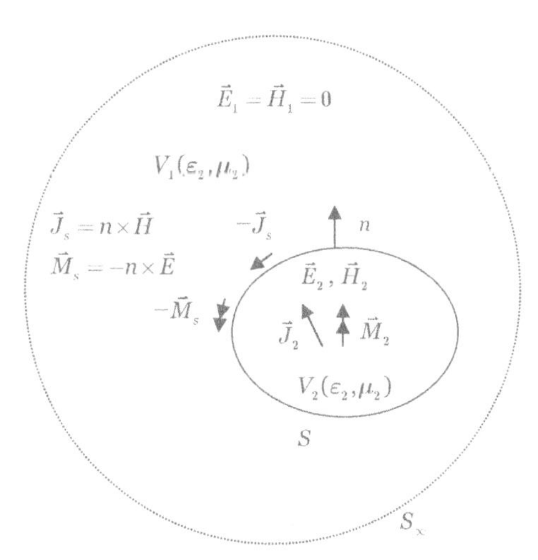

Fig. 7 When the source currents $(\vec{J}_1, \vec{M}_1)$ in V_1 are removed and the medium of region 1 is replaced with that of region 2, the source currents $(\vec{J}_2, \vec{M}_2)$ in V_2 and the negative equivalent surface currents $(-\vec{J}_1, -\vec{M}_1)$ on S will maintain the correct EM field $(\vec{E}_2, \vec{H}_2)$ in V_2 and zero EM field $\left(\vec{E}_1 = \vec{H}_1 = 0\right)$ in V_1.

1.d Applications

Mathematical formulation of the equivalence principle derived in the preceding section may have many applications.

A finite homogeneous body of arbitrary shape with complex permittivity and permeability of (ε, μ) located in space is exposed to an impressed EM field with an electric field $\vec{E}^{in}$ and a magnitude field $\vec{H}^{in}$. We aim to determine the induced EM field inside the body.

To solve the problem, we will first derive two integral equations for the equivalent surface currents, $\vec{J}_s = \hat{n} \times \vec{H}$ and $\vec{M}_s = -\hat{n} \times \vec{E}$, on the body surface in terms of $\vec{E}^{in}$ and $\vec{H}^{in}$. After solving for $\vec{J}_s$ and $\vec{M}_s$, the induced EM field inside the body can be easily calculated.

Let us consider the geometry depicted in Fig.8, a dielectric sphere with a radius a is illuminated by a plane EM wave with an electric field $\vec{E}^{in}$ and a magnitude field $\vec{H}^{in}$ incident upon the sphere from the direction of $\theta = \pi$. We can consider the sphere as region 2, V_2, and the rest of the space as region 1, V_1. Obviously, the incident EM wave is maintained by the sources $(\vec{J}_1, \vec{M}_1)$ located somewhere far away from the sphere in V_1. Also, the sources $(\vec{J}_2, \vec{M}_2)$ do not exist inside sphere, V_2.

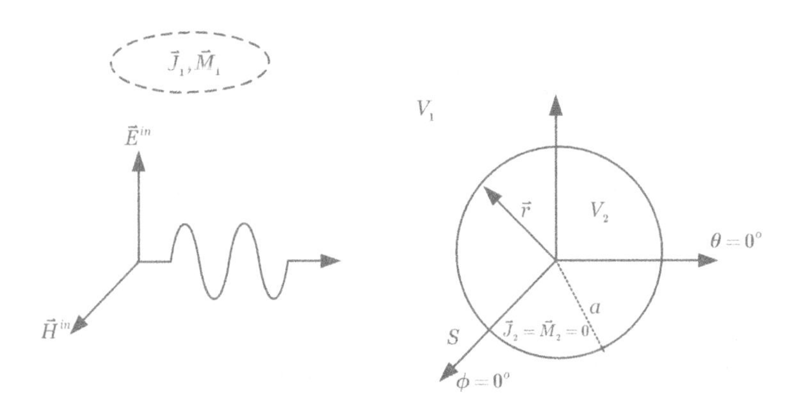

Fig. 8 A dielectric sphere is illuminated by a plane EM wave with an electric field $\vec{E}^{in}$ and a magnetic field $\vec{H}^{in}$ incident upon the sphere from the direction of $\theta = \pi$.

We now aim to find the $\vec{E}$ field at a point $\vec{r}$ on the spherical surface S in region 1 side as

$$\vec{E}_1(\vec{r}) = \frac{1}{2\pi}\int_{V_1}[-j\omega\mu_1\vec{J}_1\phi_1 - \vec{M}_1\times\nabla'\phi_1 + \frac{\rho_1}{\varepsilon_1}\nabla'\phi_1]dv' + \frac{1}{2\pi}\int_s[-j\omega\mu_1\vec{J}_s\phi_1 - \vec{M}_s\times\nabla'\phi_1 + \frac{\rho_s}{\varepsilon_1}\nabla'\phi_1]ds' \quad (1.47)$$

The volume integral of the above equation can be easily identified as twice the impressed electric field at the body surface maintained by ($\vec{J}_1$, $\vec{M}_1$), or it is equal to $2\vec{E}^{in}(\vec{r})$. Thus,

$$\vec{E}_1(\vec{r}) = 2\vec{E}^{in}(\vec{r}) + \frac{1}{2\pi}\int_s[-j\omega\mu_1\vec{J}_s\phi_1 - \vec{M}_s\times\nabla'\phi_1 + \frac{\rho_s}{\varepsilon_1}\nabla'\phi_1]ds' \quad (1.71)$$

Next, we can express the $\vec{E}$ field at the same point $\vec{r}$ on S but in region 2 side using Eq.(1.57) of the preceding section as

$$\vec{E}_2(\vec{r}) = \frac{1}{2\pi}\int_s[-j\omega\mu_2(-\vec{J}_s)\phi_2 - (-\vec{M}_s)\times\nabla'\phi_2 + \frac{-\rho_s}{\varepsilon_2}\nabla'\phi_2]ds' \quad (1.72)$$

because $\vec{J}_2 = \vec{M}_2 = 0$ in region 2.

Since the tangential component of the $\vec{E}$ field is continuous across S, or $\hat{n}\times\vec{E}_1 = n\times\vec{E}_2$, we can obtain from Eqs.(1.71) and (1.72) an integral equation as

$$\hat{n}\times\int_s[-j\omega\vec{J}_s(\mu_2\phi_2 + \mu_1\phi_1) + \vec{M}_s\times\nabla'(\phi_2+\phi_1) - \rho_s\nabla'\left(\frac{\phi_2}{\varepsilon_2} + \frac{\phi_1}{\varepsilon_1}\right)]ds' = 4\pi\times\vec{E}^{in}(\vec{r}) \quad (1.73)$$

Similarly, from the continuity of the tangential components of the $\vec{H}$ field across S, or $\hat{n}\times\vec{H}_1 = \hat{n}\times\vec{H}_2$, we can derive another integral equation as

$$\hat{n}\times\int_s[-j\omega\vec{M}_s(\varepsilon_2\phi_2 + \varepsilon_1\phi_1) - \vec{J}_s\times\nabla'(\phi_2+\phi_1) - \rho_{ms}\nabla'\left(\frac{\phi_2}{\mu_2} + \frac{\phi_1}{\mu_1}\right)]ds' = 4\pi\hat{n}\times\vec{H}^{in}(\vec{r}) \quad (1.74)$$

Eqs.(1.73) and (1.74) can be numerically solved to determine $\vec{J}_s$ and $\vec{M}_s$ by using the method of moments and vector basis functions with triangular patch modeling [3]. After $\vec{J}_s$ and $\vec{M}_s$ are determined, the $\vec{E}$ field and the $\vec{H}$ field inside the sphere can be easily computed by using Eq.(1.56) of the preceding section.

As a numerical example, the equivalent electric and magnetic surface current, $\vec{J}_s$ and $\vec{M}_s$, induced by a plane EM wave on the surface of a dielectric sphere have been computed based on Eqs.(1.73) and (1.74) and the results are shown in Figs.9 and 10. The electrical size of the sphere is $\beta_1 a=1$, where β_1 is the free-space propagation constant and a is the radius of the sphere. The permittivity of the sphere is $\varepsilon_2=4\varepsilon_0$ and the permeability is $\mu_2=\mu_0$. The plane EM wave is incident upon the sphere from the direction of $\theta=\pi$. The induced electric and magnetic currents along a circumferential arc on $\phi=0$ are plotted as function of θ in Fig.9 and 10.

Along the arc, there are two components of the electric surface current, $\vec{J}_{s\theta}$ and $\vec{J}_{s\phi}$, and two components of the magnetic surface current, $\vec{M}_{s\theta}$ and $\vec{M}_{s\phi}$. The values of $\vec{J}_s$ are shown normalized to the incident magnetic field $\vec{H}^{in}$ and those of $\vec{M}_s$ are normalized by the incident electric field $\vec{E}^{in}$.

To verify the accuracy of the numerical results, they are compared with the exact solutions of Mie series, which will be presented in the next chapter. The numerical results are indicated by small triangles and exact solutions are plotted in solid lines in Figs. 9 and 10. It is observed that very accurate numerical results can be obtained using the method based on the equivalence principle and integral equations for the induced equivalent surface currents.

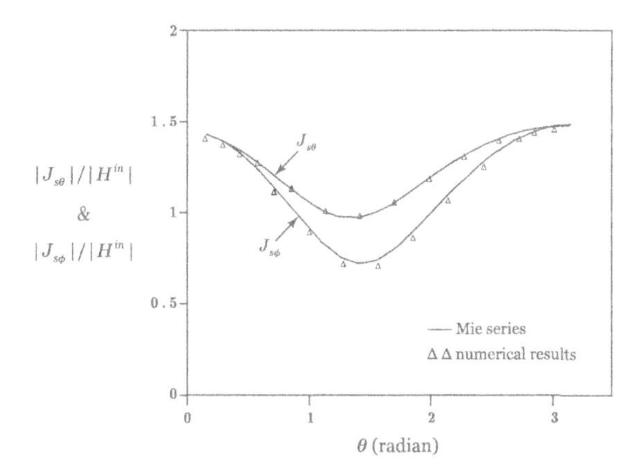

Fig. 9 Equivalent electric surface currents induced by a plane EM wave on the surface of a dielectric sphere with $\beta_1 a=1$, $\varepsilon=4\varepsilon_0$, and $\mu=\mu_0$. The plane EM wave is incident upon the sphere from the direction of $\theta=\pi$ and the surface currents are on a circumferential arc of $\phi=0$.

Reprinted from "A mathematical formulation of the equivalence principle" by Kun-Mu Chen, IEEE Transactions on Microwave Theory and Techniques, Vol. 37, No. 10, pp. 1576-1581, Oct. 1989. (©1989 IEEE.)

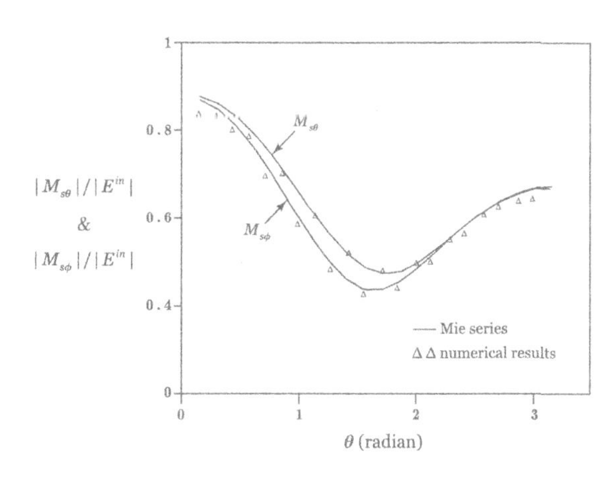

Fig. 10 Equivalent magnetic surface currents induced by a plane EM wave on the surface of a dielectric sphere with $\beta_1 a=1$, $\varepsilon=4\varepsilon_0$, and $\mu=\mu_0$. The plane EM wave is incident upon the sphere from the direction of $\theta=\pi$ and the surface currents are on a circumferential arc of $\phi=0$.

Reprinted from "A mathematical formulation of the equivalence principle" by Kun-Mu Chen, IEEE Transactions on Microwave Theory and Techniques, Vol. 37, No. 10, pp. 1576-1581, Oct. 1989. (©1989 IEEE.)

REFERENCES

[1] R.F. Harrington, "Time-Harmonic Electromagnetic Fields", New-York; McGraw-Hill, 1961, pp.106-110.

[2] K.M. Chen, "A Mathematical formulation of Equivalence Principle", *IEEE Transactions on Microwave Theory and Techniques*, vol.37, NO.10, pp 1576-1581, Oct. 1989.

[3] S.M.Rao, D.R.Wilton and A.W.Glisson, "Electromagnetic scattering by surface of arbitrary sphere", *IEEE Trans. Antennas Propagation*, vol. AP-30, pp.409-418, May 1982.

Chapter 2
Vector Wave Functions And Applications

In this chapter, we will develop vector wave functions in various coordinate systems, and use them to determine the interaction of the electromagnetic fields with material bodies.

2.a Wave Equations

Consider a region filled with a simple medium and all the EM sources are outside of the region. The electric field $\vec{E}$, the magnetic field $\vec{H}$ and the vector potential $\vec{A}$ in this region satisfy the same wave equation as shown below.

Maxwell equations for this source-free region are

$$\begin{cases} \nabla\times\vec{E} = -j\omega\mu\vec{H} & (2.1) \\ \nabla\times\vec{H} = j\omega\varepsilon\vec{E} & (2.2) \end{cases}$$

Using Eqs. (2.1) and (2.2) as

$$\nabla\times\nabla\times\vec{E} = -j\omega\mu\nabla\times\vec{H} = (-j\omega\mu)(j\omega\varepsilon)\vec{E} = \omega^2\mu\varepsilon\vec{E} = k^2\vec{E}$$

or

$$\nabla\times\nabla\times\vec{E} - k^2\vec{E} = 0 \tag{2.3}$$

since

$$\nabla^2\vec{E} = \nabla(\nabla\cdot\vec{E}) - \nabla\times\nabla\times\vec{E} = -\nabla\times\nabla\times\vec{E}$$

because

$$\nabla\cdot\vec{E} = 0 \qquad \text{from Eq. (2.2)}$$

Eq. (2.3) can be rewritten as

$$\nabla^2\vec{E}+k^2\vec{E}=0 \tag{2.4}$$

The $\vec{H}$ field satisfies the same wave equation of

$$\nabla^2\vec{H}+k^2\vec{H}=0$$

If the vector potential $\vec{A}$ is defined in this region, it should satisfy

$$\nabla^2\vec{A}+k^2\vec{A}=-\mu\vec{J}=0 \tag{2.5}$$

because $\vec{J}$ is located outside of the region.

It is noted that

$\nabla\cdot\vec{A}\neq 0$ in this case because of Lorentz condition of $\nabla\cdot\vec{A}=-j\omega\mu\varepsilon\phi$.

It is clear that $\vec{E},\vec{H}$ and $\vec{A}$ all satisfy the wave equation of

$$\nabla^2\vec{C}+k^2\vec{C}=0 \tag{2.6}$$

or

$$\nabla(\nabla\cdot\vec{C})-\nabla\times\nabla\times\vec{C}+k^2\vec{C}=0 \tag{2.7}$$

even though $\nabla\cdot\vec{E}=0$, $\nabla\cdot\vec{H}=0$ and $\nabla\cdot\vec{A}\neq 0$.

Our objective is to seek vector solution for the wave equations given in Eq. (2.6) or Eq. (2.7). We will consider the spherical coordinate system first. In order to find the vector wave function we need to start from the scalar wave function for the system first.

2.b Spherical Scalar Wave Functions

We will solve the corresponding scalar wave equation first, with the hope of constructing the appropriate vector solution for the vector wave equation based on the solution of the corresponding scalar wave equation.

The scalar wave equation corresponding to the vector wave equation of Eq. (2.6) is

$$\nabla^2 f + k^2 f = 0 \tag{2.8}$$

where $$\nabla^2 f \equiv \nabla \cdot (\nabla f)$$

If we write out $\nabla^2 f$ in spherical coordinates, we have Eq. (2.8) as

$$\frac{1}{R^2}\frac{\partial}{\partial R}(R^2\frac{\partial f}{\partial R}) + \frac{1}{R^2 \sin\theta}\frac{\partial}{\partial \theta}(\sin\theta\frac{\partial f}{\partial \theta}) + \frac{1}{R^2\sin^2\theta}\frac{\partial^2 f}{\partial \phi^2} + k^2 f = 0 \tag{2.9}$$

We will solve Eq. (2.9) by the separation of variables. Let's assume that

$$f(R,\theta,\phi) = f_1(R) f_2(\theta) f_3(\phi) \tag{2.10}$$

Substituting Eq. (2.10) in Eq. (2.9), and after dividing by f, we have

$$\frac{1}{f_1(R)}\frac{1}{R^2}\frac{\partial}{\partial R}(R^2\frac{\partial f_1(R)}{\partial R}) + \frac{1}{f_2(\theta)}\frac{1}{R^2\sin\theta}\frac{\partial}{\partial\theta}(\sin\theta\frac{\partial f_2(\theta)}{\partial\theta}) + \frac{1}{f_3(\phi)}\frac{1}{R^2\sin^2\theta}\frac{\partial^2 f_3(\phi)}{\partial\phi^2} + k^2 = 0$$

multiplying by $R^2\sin^2\theta$,

$$\frac{\sin^2\theta}{f_1(R)}\frac{\partial}{\partial R}(R^2\frac{\partial f_1(R)}{\partial R}) + \frac{\sin\theta}{f_2(\theta)}\frac{\partial}{\partial\theta}(\sin\theta\frac{\partial f_2(\theta)}{\partial\theta}) + \underbrace{\frac{1}{f_3(\phi)}\frac{\partial^2 f_3(\phi)}{\partial\phi^2}}_{\text{Functions of }\phi\text{ only}} + k^2R^2\sin^2\theta = 0$$

$$\frac{\sin^2\theta}{f_1(R)}\frac{\partial}{\partial R}\underbrace{(R^2\frac{\partial f_1(R)}{\partial R}) + \frac{\sin\theta}{f_2(\theta)}\frac{\partial}{\partial\theta}(\sin\theta\frac{\partial f_2(\theta)}{\partial\theta})}_{\text{Functions of }R\text{ and }\theta\text{ only}} + k^2R^2\sin^2\theta = -\underbrace{\frac{1}{f_3(\phi)}\frac{\partial^2 f_3(\phi)}{\partial\phi^2}}_{\text{function of }\phi\text{ only}}$$

Since ϕ and (R,θ) are independent variables, we can assume that

$$-\frac{1}{f_3(\phi)}\frac{\partial^2 f_3(\phi)}{\partial\phi^2}=m^2=\text{constant} \tag{2.11}$$

and

$$\frac{\sin^2\theta}{f_1(R)}\frac{\partial}{\partial R}(R^2\frac{\partial f_1(R)}{\partial R})+\frac{\sin\theta}{f_2(\theta)}\frac{\partial}{\partial\theta}(\sin\theta\frac{\partial f_2(\theta)}{\partial\theta})+k^2R^2\sin^2\theta=m^2=\text{same constant} \tag{2.12}$$

Eq. (2.11) gives

$$\frac{d^2 f_3(\phi)}{d\phi^2}+m^2 f_3(\phi)=0 \tag{2.13}$$

Eq. (2.13) is the harmonic equation. Eq. (2.12) leads to

$$\frac{1}{f_1(R)}\frac{d}{dR}(R^2\frac{df_1(R)}{dR})+\frac{1}{f_2(\theta)}\frac{1}{\sin\theta}\frac{\partial}{\partial\theta}(\sin\theta\frac{df_2(\theta)}{d\theta})-\frac{m^2}{\sin^2\theta}+k^2R^2=0$$

or

$$\underbrace{\frac{1}{f_1(R)}\frac{d}{dR}(R^2\frac{df_1(R)}{dR})+k^2R^2}_{\text{Functions of } R \text{ only}}=\underbrace{-\frac{1}{f_2(\theta)}\frac{1}{\sin\theta}\frac{d}{d\theta}(\sin\theta\frac{df_2(\theta)}{d\theta})+\frac{m^2}{\sin^2\theta}}_{\text{Functions of } \theta \text{ only}} \tag{2.14}$$

Since R and θ are independent variables, we can assume that

$$-\frac{1}{f_2(\theta)}\frac{1}{\sin\theta}\frac{d}{d\theta}(\sin\theta\frac{df_2(\theta)}{d\theta})+\frac{m^2}{\sin^2\theta}=p^2=n(n+1)=\text{constant}$$

Theoretically, p^2 can be any constant, however, only when $p^2 = n(n+1)$ with $n = 0,1,2,3......$ $(i.e\ \ p^2 = 0,2,6,12,.....)$, $f_2(\theta)$ can be represented by a finite-term power series and $f_2(\theta)$ will converge at $\theta = 0$ and π. Thus, we have

$$\frac{1}{\sin\theta}\frac{d}{d\theta}(\sin\theta\frac{df_2(\theta)}{d\theta}) + \left[n(n+1) - \frac{m^2}{\sin^2\theta}\right]f_2(\theta) = 0 \tag{2.15}$$

If we let $\eta = \cos\theta,\ d\eta = -\sin\theta d\theta,\ \frac{d}{d\theta} = -\sin\theta\frac{1}{d\eta},$

$$\frac{1}{\sin\theta}\frac{d}{d\theta}(\sin\theta\frac{df_2(\theta)}{d\theta}) = \frac{1}{d\eta}((\sin^2\theta)\frac{df_2(\eta)}{d\eta}) = \sin^2\theta\frac{d^2 f_2(\eta)}{d\eta^2} + 2\sin\theta cos\theta\frac{d\theta}{d\eta}\frac{df_2(\eta)}{d\eta}$$

$$= \left(1-\eta^2\right)\frac{d^2 f_2(\eta)}{d\eta^2} - 2\eta\frac{df_2(\eta)}{d\eta}$$

Hence, Eq. (2.15) becomes

$$\left(1-\eta^2\right)\frac{d^2 f_2(\eta)}{d\eta^2} - 2\eta\frac{df_2(\eta)}{d\eta} + \left[n(n+1) - \frac{m^2}{1-\eta^2}\right]f_2(\eta) = 0 \tag{2.16}$$

Eq. (2.16) is the ***Legendre*** equation.

Going bach to Eq. (2.14), we should set the L.H.S of Eq. (2.14) equal to the same constant, $n(n+1)$:

$$\frac{1}{f_1(R)}\frac{d}{dR}(R^2\frac{df_1(R)}{dR}) + k^2R^2 = n(n+1)$$

or

$$\frac{d}{dR}(R^2\frac{df_1(R)}{dR}) + \left[k^2R^2 - n(n+1)\right]f_1(R) = 0$$

$$R^2\frac{d^2 f_1(R)}{dR^2} + 2R\frac{df_1(R)}{dR} + \left[k^2R^2 - n(n+1)\right]f_1(R) = 0 \tag{2.17}$$

If we let

$$f_1(R) = \frac{1}{\sqrt{kR}} V(R),$$

We can transform Eq. (2.17) into

$$R^2 \frac{d^2V(R)}{dR^2} + R\frac{dV(R)}{dR} + \left[k^2R^2 - (n+\frac{1}{2})^2\right]V(R) = 0 \tag{2.18}$$

Eq. (2.18) is the ***Bessel*** function of order $(n+\frac{1}{2})$.

The solution to Eq. (2.13) is

$$f_3(\phi) = \begin{Bmatrix} \cos m\phi \\ \sin m\phi \end{Bmatrix} \tag{2.19}$$

The solution to Eq. (2.16) is

$$f_2(\theta) = f_2(\eta) = P_n^m(\eta) = P_n^m(\cos\theta) \tag{2.20}$$

where $P_n^m(\cos\theta)$ is the associated ***Legendre*** polynomials.

The solution to Eq. (2.18) is

$$V(R) = Z_{n+\frac{1}{2}}(kR) \tag{2.21}$$

where $Z_{n+\frac{1}{2}}(kR)$ is any ***Bessel*** function of fractional order, $n+\frac{1}{2}$.

That is

$$Z_{n+\frac{1}{2}}(kR)=\begin{Bmatrix} J_{n+\frac{1}{2}}(kR) \\ Y_{n+\frac{1}{2}}(kR) \end{Bmatrix} \ or \ \begin{Bmatrix} H^{(1)}_{n+\frac{1}{2}}(kR) \\ H^{(2)}_{n+\frac{1}{2}}(kR) \end{Bmatrix} \tag{2.22}$$

The solution for $f_1(R)$ is then

$$f_1(R)=\frac{1}{\sqrt{kR}}Z_{n+\frac{1}{2}}(kR)\equiv\sqrt{\frac{2}{\pi}}z_n(kR) \tag{2.23}$$

where $z_n(kR)$ is the Spherical ***Bessel*** function.

Thus, the final solution for $f(R,\theta,\phi)$ is

$$f(R,\theta,\phi)=f_1(R)f_2(\theta)f_3(\phi)=\frac{1}{\sqrt{kR}}Z_{n+\frac{1}{2}}(kR)P_n^m(\cos\theta)\begin{Bmatrix} \cos m\phi \\ \sin m\phi \end{Bmatrix} \tag{2.24}$$

or, in an often-used notation,

$$f_{{}^e_omn}=z_n(kR)P_n^m(\cos\theta)\begin{Bmatrix} \cos m\phi \\ \sin m\phi \end{Bmatrix} \qquad \begin{matrix} e\rightarrow even(\cos m\phi) \\ o\rightarrow odd(\sin m\phi) \\ \text{m and n can be any integer} \end{matrix} \tag{2.25}$$

2.c Spherical Vector Wave Functions

Now that we have found the solution to the scalar wave equation, we will construct the solution to the vector wave equation based on the former.

We are solving for Eq. (2.6) and Eq. (2.7) :

$$\nabla^2\vec{C}+k^2\vec{C}=0 \tag{2.6}$$

or

$$\nabla(\nabla\cdot\vec{C})-\nabla\times\nabla\times\vec{C}+k^2\vec{C}=0 \tag{2.7}$$

We can show that there exist three possible solutions to Eq. (2.6) and Eq. (2.7). They are

$$\vec{L}(R,\theta,\phi)=\nabla f(R,\theta,\phi) \tag{2.26}$$

$$\vec{M}(R,\theta,\phi)=\nabla\times\left[\vec{R}f(R,\theta,\phi)\right] \tag{2.27}$$

$$(\vec{R}=\hat{R}R)$$

$$\vec{N}(R,\theta,\phi)=\frac{1}{k}\nabla\times\vec{M}(R,\theta,\phi) \tag{2.28}$$

where $f(R,\theta,\phi)$ is the solution of the scalar wave equation as given in Eq. (2.25). We will prove $\vec{L}$, $\vec{M}$ and $\vec{N}$ functions to be the solutions of Eq. (2.7).

Proof:

(1) Substituting $\vec{L}=\nabla f$ in the L.H.S of Eq. (2.7) leads to

$$\nabla(\nabla\cdot\nabla f)-\nabla\times\underbrace{\nabla\times(\nabla f)}_{=0}+k^2\nabla f$$

$$=\nabla(\nabla^2 f)+k^2\nabla f=\nabla\left[\nabla^2 f+k^2 f\right]=0$$

because $\nabla^2 f+k^2 f=0$

Therefore, $\vec{L}$ is a solution of Eq. (2.7) or Eq. (2.6). It is noted that

$$\begin{cases} \nabla\times\vec{L}=0 \\ \nabla\cdot\vec{L}\neq 0 \end{cases} \tag{2.29}$$

Also, even though $\vec{L}$ is a solution for Eq. (2.7), it is not an appropriate solution for $\vec{E}$ or $\vec{H}$ because $\nabla\cdot\vec{E}=0$ and $\nabla\cdot\vec{H}=0$. $\vec{L}$ can be used to represent a longitudinal field which has the properties of Eq. (2.29).

(2) To prove that $\vec{M}=\nabla\times(\vec{R}f)$ is a solution of Eq. (2.7), it requires a tedious manipulation.

Substituting $\vec{M}$ in the L.H.S of Eq. (2.7) gives

$$\nabla\underbrace{\left[\nabla\cdot\nabla\times(\vec{R}f)\right]}_{=0}-\nabla\times\nabla\times\left[\nabla\times(\vec{R}f)\right]+k^2\nabla\times(\vec{R}f)$$

$$=-\nabla\times\nabla\times\left[\nabla\times(\vec{R}f)\right]+k^2\nabla\times(\vec{R}f)$$

In spherical coordinates,

$$\begin{aligned}\nabla\times\vec{A}=&\frac{1}{R\sin\theta}\left[\frac{\partial}{\partial\theta}(\sin\theta A_\phi)-\frac{\partial A_\theta}{\partial\phi}\right]\ddot{R}\\&+\frac{1}{R}\left[\frac{1}{\sin\theta}\frac{\partial A_R}{\partial\phi}-\frac{\partial}{\partial R}(RA_\phi)\right]\hat{\theta}\\&+\frac{1}{R}\left[\frac{\partial}{\partial R}(RA_\theta)-\frac{\partial A_R}{\partial\theta}\right]\hat{\phi}\end{aligned}$$

then

$$\nabla\times(\vec{R}f)=\nabla\times(\hat{R}Rf)=\frac{1}{R}\left[\frac{1}{\sin\theta}\frac{\partial}{\partial\phi}(Rf)\right]\hat{\theta}+\frac{1}{R}\left[-\frac{\partial}{\partial\theta}(Rf)\right]\hat{\phi}$$

$$=\left[\frac{1}{\sin\theta}\frac{\partial f}{\partial\phi}\right]\hat{\theta}+\left[-\frac{\partial f}{\partial\theta}\right]\hat{\phi}=\vec{M}$$

$$\nabla\times\left[\nabla\times(\vec{R}f)\right]=\frac{1}{R\sin\theta}\left[\frac{\partial}{\partial\theta}(-\sin\theta\frac{\partial f}{\partial\theta})-\frac{\partial}{\partial\phi}(\frac{1}{\sin\theta}\frac{\partial f}{\partial\phi})\right]\hat{R}$$

$$+\frac{1}{R}\left[-\frac{\partial}{\partial R}(-R\frac{\partial f}{\partial\theta})\right]\hat{\theta}+\frac{1}{R}\left[\frac{\partial}{\partial R}(R\frac{1}{\sin\theta}\frac{\partial f}{\partial\phi})\right]\hat{\phi}$$

$$=\frac{1}{R\sin\theta}\left[-\frac{\partial}{\partial\theta}(\sin\theta\frac{\partial f}{\partial\theta})-\frac{1}{\sin\theta}\frac{\partial^2 f}{\partial\phi^2}\right]\hat{R}$$

$$+\left[\frac{1}{R}\frac{\partial}{\partial R}(R\frac{\partial f}{\partial\theta})\right]\hat{\theta}+\frac{1}{R\sin\theta}\left[\frac{\partial}{\partial R}(R\frac{\partial f}{\partial\phi})\right]\hat{\phi}$$

$$\nabla\times\nabla\times\left[\nabla\times(\vec{R}f)\right]=\frac{1}{R\sin\theta}\left\{\frac{\partial}{\partial\theta}\left[\sin\theta\frac{1}{R\sin\theta}\frac{\partial}{\partial R}(R\frac{\partial f}{\partial\phi})\right]-\frac{\partial}{\partial\phi}\left[\frac{1}{R}\frac{\partial}{\partial R}(R\frac{\partial f}{\partial\theta})\right]\right\}\hat{R}$$

$$+\frac{1}{R}\left\{\frac{1}{\sin\theta}\frac{1}{R\sin\theta}\left[-\frac{\partial}{\partial\theta}(\sin\theta\frac{\partial^2 f}{\partial\phi\partial\theta})-\frac{1}{\sin\theta}\frac{\partial^3 f}{\partial\phi^3}\right]-\frac{\partial}{\partial R}\left[\frac{1}{\sin\theta}\frac{\partial}{\partial R}(R\frac{\partial f}{\partial\phi})\right]\right\}\hat{\theta}$$

$$+\frac{1}{R}\left\{\frac{\partial}{\partial R}\left[\frac{\partial}{\partial R}(R\frac{\partial f}{\partial\theta})\right]-\frac{\partial}{\partial\theta}\left[\frac{1}{R\sin\theta}\left[-\frac{\partial}{\partial\theta}(\sin\theta\frac{\partial f}{\partial\theta})-\frac{1}{\sin\theta}\frac{\partial^2 f}{\partial\phi^2}\right]\right]\right\}\hat{\phi}$$

$$=\frac{1}{R\sin\theta}\left\{\underbrace{\frac{\partial}{\partial\theta}\left[\frac{1}{R}\frac{\partial}{\partial R}(R\frac{\partial f}{\partial\phi})\right]-\frac{\partial}{\partial\theta}\left[\frac{1}{R}\frac{\partial}{\partial R}(R\frac{\partial f}{\partial\phi})\right]}_{0}\right\}\hat{R}$$

$$+\left\{\frac{1}{R^2\sin^2\theta}\left[-\frac{\partial}{\partial\theta}(\sin\theta\frac{\partial^2 f}{\partial\phi\partial\theta})-\frac{1}{\sin\theta}\frac{\partial^3 f}{\partial\phi^3}\right]-\frac{1}{R\sin\theta}\frac{\partial^2}{\partial R^2}(R\frac{\partial f}{\partial\phi})\right\}\hat{\theta}$$

$$+\frac{1}{R}\left\{\frac{\partial^2}{\partial R^2}(R\frac{\partial f}{\partial\theta})+\frac{1}{R}\frac{\partial}{\partial\theta}\left[\frac{1}{\sin\theta}\frac{\partial}{\partial\theta}(\sin\theta\frac{\partial f}{\partial\theta})\right]+\frac{1}{R}\frac{\partial}{\partial\theta}(\frac{1}{\sin^2\theta}\frac{\partial^2 f}{\partial\phi^2})\right\}\hat{\phi}$$

Now, the L.H.S of Eq. (2.7) becomes

$$-\nabla\times\nabla\times\left[\nabla\times(\vec{R}f)\right]+k^2\nabla\times(\vec{R}f)$$

$$=\left\{\frac{1}{R^2\sin^2\theta}\left[\frac{\partial}{\partial\theta}(\sin\theta\frac{\partial^2 f}{\partial\phi\partial\theta})+\frac{1}{\sin\theta}\frac{\partial^3 f}{\partial\phi^3}\right]+\frac{1}{R\sin\theta}\frac{\partial^2}{\partial R^2}(R\frac{\partial f}{\partial\phi})+\frac{k^2}{\sin\theta}\frac{\partial f}{\partial\phi}\right\}\hat{\theta}$$

$$+\frac{-1}{R}\left\{\frac{\partial^2}{\partial R^2}(R\frac{\partial f}{\partial\theta})+\frac{1}{R}\frac{\partial}{\partial\theta}\left[\frac{1}{\sin\theta}\frac{\partial}{\partial\theta}(\sin\theta\frac{\partial f}{\partial\theta})\right]+\frac{1}{R}\frac{\partial}{\partial\theta}(\frac{1}{\sin^2\theta}\frac{\partial^2 f}{\partial\phi^2})+k^2R\frac{\partial f}{\partial\theta}\right\}\hat{\phi}$$

$$=\frac{1}{\sin\theta}\frac{\partial}{\partial\phi}\left[\underbrace{\underbrace{\frac{1}{R}\frac{\partial^2}{\partial R^2}(Rf)}_{\frac{1}{R^2}\frac{\partial}{\partial R}(R^2\frac{\partial f}{\partial R})}+\frac{1}{R^2\sin\theta}\frac{\partial}{\partial\theta}(\sin\theta\frac{\partial f}{\partial\theta})+\frac{1}{R^2\sin^2\theta}\frac{\partial^2 f}{\partial\phi^2}+k^2 f}_{=0\text{ due to Eq.(2.9)}}\right]\hat{\theta}$$

$$-\frac{\partial}{\partial\theta}\left[\underbrace{\frac{1}{R}\frac{\partial^2}{\partial R^2}(Rf)+\frac{1}{R^2\sin\theta}\frac{\partial}{\partial\theta}(\sin\theta\frac{\partial f}{\partial\theta})+\frac{1}{R^2\sin^2\theta}\frac{\partial^2 f}{\partial\phi^2}+k^2 f}_{=0\text{ due to Eq.(2.9)}}\right]\hat{\phi}$$

$$=0$$

Therefore, $\vec{M}=\nabla\times(\vec{R}f)$ is a solution to Eq. (2.7). It is noted that

$$\begin{cases}\nabla\cdot\vec{M}=0\\ \nabla\times\vec{M}\neq 0\end{cases}\qquad(2.30)$$

Thus, $\vec{M}$ is an appropriate solution for $\vec{E}$ and $\vec{H}$.

(3) It is an easy matter to prove that $\vec{N} = \frac{1}{R}\nabla\times\vec{M}$ is another solution of Eq. (2.7). Substituting $\vec{N}$ in the L.H.S of Eq. (2.7), we have

$$\nabla(\nabla\cdot\vec{N}) - \nabla\times\nabla\times\vec{N} + k^2\vec{N}$$

$$= \nabla\underbrace{\left[\nabla\cdot(\frac{1}{k}\nabla\times\vec{M})\right]}_{=0} - \nabla\times\nabla\times(\frac{1}{k}\nabla\times\vec{M}) + k^2(\frac{1}{k}\nabla\times\vec{M})$$

$$= -\frac{1}{k}\nabla\times\nabla\times(\nabla\times\vec{M}) + k\nabla\times\vec{M}$$

Since we have proved that

$$-\nabla\times\nabla\times\vec{M} + k^2\vec{M} = 0$$

$$-\frac{1}{k}\nabla\times(k^2\vec{M}) + k\nabla\times\vec{M} = 0$$

Therefore, $\vec{N}$ is another solution to Eq. (2.7). It is also noted that

$$\begin{cases} \nabla\cdot\vec{N} = 0 \\ \nabla\times\vec{N} \neq 0 \end{cases} \tag{2.31}$$

So that $\vec{N}$ is also an appropriate solution for $\vec{E}$ or $\vec{H}$ field.

We have now proved that $\vec{L}$, $\vec{M}$ and $\vec{N}$ are three possible vector solutions to the wave equation of Eq. (2.6) or Eq. (2.7). We will determine the complete expressions for those three vector solutions.

$$\vec{L}(R,\theta,\phi) = \nabla f(R,\theta,\phi)$$

Since

$$\nabla f = \frac{\partial f}{\partial R}\hat{R} + \frac{1}{R}\frac{\partial f}{\partial\theta}\hat{\theta} + \frac{1}{R\sin\theta}\frac{\partial f}{\partial\phi}\hat{\phi}$$

$$f = f_{\substack{e\\o}mn} = z_n(kR)P_n^m(\cos\theta)\begin{Bmatrix}\cos m\phi \\ \sin m\phi\end{Bmatrix}$$

we can write down $\vec{L}$ as

$$\vec{L}_{\substack{e\\o}mn} = \left[\frac{\partial}{\partial R} z_n(kR)\right] P_n^m(\cos\theta) \begin{Bmatrix} \cos m\phi \\ \sin m\phi \end{Bmatrix} \hat{R}$$

$$+ \frac{1}{R} z_n(kR) \left[\frac{\partial}{\partial \theta} P_n^m(\cos\theta)\right] \begin{Bmatrix} \cos m\phi \\ \sin m\phi \end{Bmatrix} \hat{\theta}$$

$$\mp \frac{m}{R\sin\theta} z_n(kR) P_n^m(\cos\theta) \begin{Bmatrix} \sin m\phi \\ \cos m\phi \end{Bmatrix} \hat{\phi} \qquad (2.32)$$

For the $\vec{M}$ function,

$$\vec{M} = \nabla \times (\vec{R} f) = \left[\frac{1}{\sin\theta} \frac{\partial f}{\partial \phi}\right] \hat{\theta} + \left[-\frac{\partial f}{\partial \theta}\right] \hat{\phi}$$ (See page 48)

we have

$$\vec{M}_{\substack{e\\o}mn} = \mp \frac{m}{\sin\theta} z_n(kR) P_n^m(\cos\theta) \begin{Bmatrix} \sin m\phi \\ \cos m\phi \end{Bmatrix} \hat{\theta}$$

$$- z_n(kR) \left[\frac{\partial}{\partial \theta} P_n^m(\cos\theta)\right] \begin{Bmatrix} \cos m\phi \\ \sin m\phi \end{Bmatrix} \hat{\phi} \qquad (2.33)$$

It is noted that $\vec{M}$ function does not have a radial component. For the $\vec{N}$ function,

$$\vec{N} = \frac{1}{k} \nabla \times \vec{M} = \frac{1}{k} \nabla \times \nabla \times (\vec{R} f)$$

From page 48, $\vec{N}$ is found to be

$$\vec{N} = \frac{1}{kR\sin\theta} \left[-\frac{\partial}{\partial \theta}(\sin\theta \frac{\partial f}{\partial \theta}) - \frac{1}{\sin\theta} \frac{\partial^2 f}{\partial \phi^2}\right] \hat{R}$$

$$+ \left[\frac{1}{kR} \frac{\partial}{\partial R}(R \frac{\partial f}{\partial \theta})\right] \hat{\theta} + \frac{1}{kR\sin\theta} \left[\frac{\partial}{\partial R}(R \frac{\partial f}{\partial \phi})\right] \hat{\phi}$$

The R -component of $\vec{N}$ can be simplified as follow :

$$\frac{1}{kR\sin\theta}\left[-\frac{\partial}{\partial\theta}(\sin\theta\frac{\partial f}{\partial\theta})-\frac{1}{\sin\theta}\frac{\partial^2 f}{\partial\phi^2}\right]$$

$$=\frac{-R}{k}\left[\frac{1}{R^2\sin\theta}\frac{\partial}{\partial\theta}(\sin\theta\frac{\partial f}{\partial\theta})+\frac{1}{R^2\sin^2\theta}\frac{\partial^2 f}{\partial\phi^2}\right]$$

$$=\frac{-R}{k}\left[-\frac{1}{R^2}\frac{\partial}{\partial R}(R^2\frac{\partial f}{\partial R})-k^2 f\right] \qquad \text{due to Eq. (2.9)}$$

$$=\frac{1}{kR}(2R\frac{\partial f}{\partial R}+R^2\frac{\partial^2 f}{\partial R^2})+kRf$$

$$=\frac{1}{kR}\left[-k^2R^2+n(n+1)\right]f+kRf \qquad \text{due to Eq. (2.17)}$$

$$=\frac{1}{kR}n(n+1)f$$

Therefore,

$$\vec{N}=\left[\frac{n(n+1)}{kR}f\right]\hat{R}+\left[\frac{1}{kR}\frac{\partial}{\partial R}(R\frac{\partial f}{\partial\theta})\right]\hat{\theta}+\frac{1}{kR\sin\theta}\left[\frac{\partial}{\partial R}(R\frac{\partial f}{\partial\phi})\right]\hat{\phi}$$

We can write down $\vec{N}$ as

$$\vec{N}_{\substack{e\\o}mn}=\frac{n(n+1)}{kR}z_n(kR)P_n^m(\cos\theta)\begin{Bmatrix}\cos m\phi\\ \sin m\phi\end{Bmatrix}\hat{R}$$

$$+\frac{1}{kR}\frac{\partial}{\partial R}[Rz_n(kR)]\left[\frac{\partial}{\partial\theta}P_n^m(\cos\theta)\right]\begin{Bmatrix}\cos m\phi\\ \sin m\phi\end{Bmatrix}\hat{\theta}$$

$$\mp\frac{m}{kR\sin\theta}\frac{\partial}{\partial R}[Rz_n(kR)]P_n^m(\cos\theta)\begin{Bmatrix}\sin m\phi\\ \cos m\phi\end{Bmatrix}\hat{\phi} \qquad (2.34)$$

We can now summarize the spherical vector wave functions as follow

$$\vec{L}_{\substack{e\\o}mn}=\left[\frac{\partial}{\partial R}z_n(kR)\right]P_n^m(\cos\theta)\begin{Bmatrix}\cos m\phi\\ \sin m\phi\end{Bmatrix}\hat{R}$$

$$+\frac{1}{R}z_n(kR)\left[\frac{\partial}{\partial\theta}P_n^m(\cos\theta)\right]\begin{Bmatrix}\cos m\phi\\ \sin m\phi\end{Bmatrix}\hat{\theta}$$

$$\mp\frac{m}{R\sin\theta}z_n(kR)P_n^m(\cos\theta)\begin{Bmatrix}\sin m\phi\\ \cos m\phi\end{Bmatrix}\hat{\phi} \tag{2.32}$$

$$\vec{M}_{\substack{e\\o}mn}=\mp\frac{m}{\sin\theta}z_n(kR)P_n^m(\cos\theta)\begin{Bmatrix}\sin m\phi\\ \cos m\phi\end{Bmatrix}\hat{\theta}$$

$$-z_n(kR)\left[\frac{\partial}{\partial\theta}P_n^m(\cos\theta)\right]\begin{Bmatrix}\cos m\phi\\ \sin m\phi\end{Bmatrix}\hat{\phi} \tag{2.33}$$

$$\vec{N}_{\substack{e\\o}mn}=\frac{n(n+1)}{\kappa R}z_n(kR)P_n^m(\cos\theta)\begin{Bmatrix}\cos m\phi\\ \sin m\phi\end{Bmatrix}\hat{R}$$

$$+\frac{1}{kR}\frac{\partial}{\partial R}[Rz_n(kR)]\left[\frac{\partial}{\partial\theta}P_n^m(\cos\theta)\right]\begin{Bmatrix}\cos m\phi\\ \sin m\phi\end{Bmatrix}\hat{\theta}$$

$$\mp\frac{m}{kR\sin\theta}\frac{\partial}{\partial R}[Rz_n(kR)]P_n^m(\cos\theta)\begin{Bmatrix}\sin m\phi\\ \cos m\phi\end{Bmatrix}\hat{\phi} \tag{2.34}$$

where

$$z_n(kR)=\frac{\sqrt{\pi/2}}{\sqrt{kR}}Z_{n+\frac{1}{2}}(kR),\ m,n=\text{any integers}$$

$$\nabla\cdot\vec{M}=0\ ,\ \nabla\cdot\vec{N}=0\ ,\ \nabla\times\vec{L}=0$$

2.d Scattering of a Conducting Sphere — Application

In this section, we will apply the spherical vector wave functions to find the scattering of conducting sphere. From the results we can find why the sky looks blue.

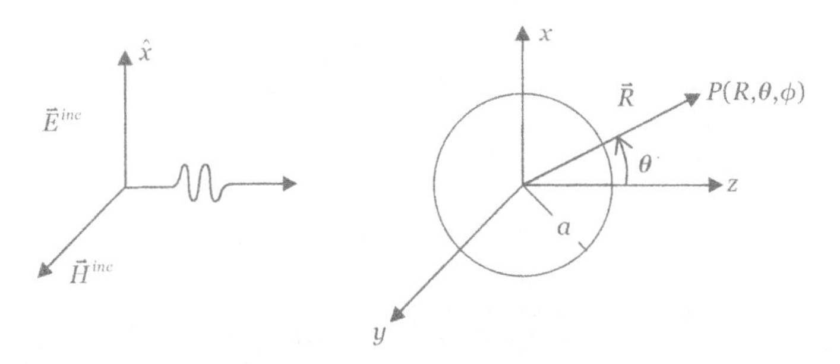

Fig. 1 A conducting sphere illuminated by a plane EM wave.

A perfectly conducting sphere of radius a is illuminated by a plane EM wave propagating in the $+z$ -direction and with an electric field of unit intensity as shown in Fig. 1.

$$\vec{E}^{inc} = \hat{x}e^{-jkz} = \hat{x}e^{-jkR\cos\theta} \tag{2.35}$$

$$\because\ z = R\cos\theta$$

$$\vec{H}^{inc} = \hat{y}\frac{1}{\zeta}e^{-jkz} = \hat{y}\frac{1}{\zeta}e^{-jkR\cos\theta} \tag{2.36}$$

When this plane EM wave is incident on the sphere, current and charge are induced on the spherical surface, and in turn, they will maintain a scattered EM wave. To determine the scattered EM wave, the incident EM wave will be expanded into spherical vector wave functions first.

The spherical components of $\hat{x}$ are

$$[\hat{x}]_R = \hat{x}\cdot\hat{R} = \hat{x}\cdot(\hat{x}\sin\theta\cos\phi + \hat{y}\sin\theta\sin\phi + \hat{z}\cos\theta) = \sin\theta\cos\phi$$

$$[\hat{x}]_\theta = \hat{x}\cdot\hat{\theta} = \hat{x}\cdot(\hat{x}\cos\theta\cos\phi + \hat{y}\cos\theta\sin\phi - \hat{z}\sin\theta) = \cos\theta\cos\phi$$

$$[\hat{x}]_\phi = \hat{x}\cdot\hat{\phi} = \hat{x}\cdot(-\hat{x}\sin\phi + \hat{y}\cos\phi) = -\sin\phi$$

Hence,

$$\hat{x} = \sin\theta\cos\phi\hat{R} + \cos\theta\cos\phi\hat{\theta} - \sin\phi\hat{\phi} \tag{2.37}$$

and

$$\vec{E}^{inc} = \left[\sin\theta\cos\phi\hat{R} + \cos\theta\cos\phi\hat{\theta} - \sin\phi\hat{\phi}\right]e^{-jkR\cos\theta} \tag{2.38}$$

Since $\nabla\cdot\vec{E}^{inc} = 0$, we need $\vec{M}$ and $\vec{N}$ functions to represent it. ($\vec{L}$ function is not appropriate here).

$\vec{N}$ has $\hat{R}$, $\hat{\theta}$ and $\hat{\phi}$ components. If we compare ϕ dependents of $\vec{E}^{inc}$ and $\vec{N}$, we conclude that we need :

$$even\ \vec{N}\ function\ and\ m = 1$$

$$\text{or } \vec{N}_{e1n}$$

For $\vec{M}$ function, it has $\hat{\theta}$ and $\hat{\phi}$ components. If we compare ϕ dependents of $\vec{E}^{inc}$ and $\vec{M}$, we conclude that we need :

$$odd\ \vec{M}\ function\ and\ m = 1$$

$$\text{or } \vec{M}_{o1n}$$

We then try to represent

$$\hat{x}e^{-jkR\cos\theta} = \left[\sin\theta\cos\phi\hat{R} + \cos\theta\cos\phi\hat{\theta} - \sin\phi\hat{\phi}\right]e^{-jkR\cos\theta}$$

$$= \sum_{n=0}^{\infty}\left[a_n\,\vec{M}_{o1n} + b_n\,\vec{N}_{e1n}\right] \tag{2.39}$$

The spherical Bessel function $z_n(kR)$ for $\vec{M}_{o1n}$ and $\vec{N}_{e1n}$ in Eq. (2.39) should be

$$z_n(kR) = j_n(kR) = \frac{\sqrt{\pi/2}}{\sqrt{kR}}J_{n+\frac{1}{2}}(kR)$$

because the incident electric field, $\hat{x}e^{-jkR\cos\theta}$, is finite at the origin $R=0$. That means that other Bessel Functions are not allowed because they diverge at $R=0$.

To determine b_n, let's compare the $\hat{R}$ components of both sides of Eq. (2.39) :

$$\sin\theta\cos\phi\, e^{-jkR\cos\theta} = \sum_{n=0}^{\infty} b_n \frac{n(n+1)}{kR} j_n(kR)P_n^1(\cos\theta)\cos\phi$$

or

$$\sin\theta\, e^{-jkR\cos\theta} = \sum_{n=1}^{\infty} b_n \frac{n(n+1)}{kR} j_n(kR)P_n^1(\cos\theta) \quad \text{for } n=0 \text{ will make R.H.S.}=0$$

Since

$$\frac{d}{d\theta}(e^{-jkR\cos\theta}) = jkR\sin\theta\, e^{-jkR\cos\theta}$$

we have

$$\frac{1}{jkR}\frac{d}{d\theta}(e^{-jkR\cos\theta}) = \sum_{n=1}^{\infty} b_n \frac{n(n+1)}{kR} j_n(kR)P_n^1(\cos\theta) \tag{2.40}$$

At this point, we need to represent $e^{-jkR\cos\theta}$ with spherical scalar wave functions, since $e^{-jkR\cos\theta}$ is not a function of ϕ or symmetrical about the polar axis, we can assume that

$$e^{-jkR\cos\theta} = \sum_{n=0}^{\infty} c_n j_n(kR) P_n(\cos\theta) \tag{2.41}$$

Multiplying both sides of Eq. (2.41) with $P_l(\cos\theta)\sin\theta$ and integrating with respect to θ from 0 to π, we have

$$\begin{aligned}\int_0^{\pi} e^{-jkR\cos\theta} P_l(\cos\theta)\sin\theta d\theta &= c_l j_l(kR) \int_0^{\pi} P_l^2(\cos\theta)\sin\theta d\theta \\ &= c_l j_l(kR)\left(\frac{2}{2l+1}\right)\end{aligned}$$

To eliminate the R dependence of the above equation, let's differentiate both sides with respect to kR l times and then set $R=0$:

$$\int_0^{\pi} (-j\cos\theta)^l e^{-jkR\cos\theta} P_l(\cos\theta)\sin\theta d\theta = C_l\left(\frac{2}{2l+1}\right)\left[\frac{d^l j_l(kR)}{d(kR)^l}\right]$$

$$(-j)^l \int_0^{\pi} \cos^l\theta\, P_l(\cos\theta)\sin\theta d\theta = C_l\left(\frac{2}{2l+1}\right)\left[\frac{d^l j_l(kR)}{d(kR)^l}\right]_{R=0}$$

$$\because \quad \left[\frac{d^l j_l(kR)}{d(kR)^l}\right]_{R=0} = \frac{2^l (l!)^2}{(2l+1)!}$$

and

$$\int_0^{\pi} \cos^l\theta\, P_l(\cos\theta)\sin\theta d\theta = \frac{2^{l+1}(l!)^2}{(2l+1)!}$$

Hence

$$(-j)^l \frac{2^{l+1}(l!)^2}{(2l+1)!} = C_l \frac{2^{l+1}(l!)^2}{(2l+1)(2l+1)!}$$

or

$$C_l = (-j)^l (2l+1)$$

Therefore, we have

$$e^{-jkR\cos\theta} = \sum_{n=0}^{\infty} (-j)^n (2n+1) J_n(kR) P_n(\cos\theta) \tag{2.42}$$

Substituting Eq. (2.42) back into Eq. (2.40), we have

$$\frac{1}{jkR} \frac{d}{d\theta} \left[\sum_{n=0}^{\infty} (-j)^n (2n+1) J_n(kR) P_n(\cos\theta) \right] = \sum_{n=1}^{\infty} b_n \frac{n(n+1)}{kR} J_n(kR) P_n^1(\cos\theta)$$

$$\because \quad \frac{d}{d\theta} P_n(\cos\theta) = -P_n^1(\cos\theta)$$

we have

$$\frac{-1}{jkR} \sum_{n=0}^{\infty} (-j)^n (2n+1) J_n(kR) P_n^1(\cos\theta) = \sum_{n=1}^{\infty} b_n \frac{n(n+1)}{kR} J_n(kR) P_n^1(\cos\theta)$$

$$\because \quad P_0^1(\cos\theta) = 0 \text{, or } \frac{d}{d\theta} P_0(\cos\theta) = \frac{d}{d\theta}(1) = 0$$

so

$$\frac{j}{kR} \sum_{n=1}^{\infty} (-j)^n (2n+1) J_n(kR) P_n^1(\cos\theta) = \frac{1}{kR} \sum_{n=1}^{\infty} b_n n(n+1) J_n(kR) P_n^1(\cos\theta)$$

That is

$$b_n = j(-j)^n \frac{2n+1}{n(n+1)} = (-j)^{n-1} \frac{2n+1}{n(n+1)} \tag{2.43}$$

To determine a_n, we will compare the $\hat{R}$ component of $\vec{H}^{inc}$ field.

From Eq. (2.36),

$$\vec{H}^{inc} = \hat{y}\frac{1}{\zeta}e^{-jkR\cos\theta} \tag{2.36}$$

Since

$$\hat{y} = \sin\theta\sin\phi\,\hat{R} + \cos\theta\sin\phi\,\hat{\theta} + \cos\phi\,\hat{\phi}$$

$$\vec{H}^{inc} = \frac{1}{\zeta}\left[\sin\theta\sin\phi\,\hat{R} + \cos\theta\sin\phi\,\hat{\theta} + \cos\phi\,\hat{\phi}\right]e^{-jkR\cos\theta} \tag{2.44}$$

From Maxwell's Eqs.

$$\vec{H}^{inc} = \frac{1}{\omega\mu}\nabla\times\vec{E}^{inc}$$

and we have represented $\vec{E}^{inc}$ with

$$\vec{E}^{inc} = \sum_{n=0}^{\infty}\left[a_n\vec{M}_{o1n} + b_n\vec{N}_{e1n}\right] \tag{2.39}$$

So that

$$\vec{H}^{inc} = \frac{j}{\omega\mu}\sum_{n=0}^{\infty}\left[a_n\nabla\times\vec{M}_{o1n} + b_n\nabla\times\vec{N}_{e1n}\right]$$

Since

$$\nabla \times \vec{M}_{o1n} = k\vec{N}_{o1n}$$

and

$$\nabla \times \vec{N}_{e1n} = \frac{1}{k}\nabla \times \nabla \times \vec{M}_{e1n} = \frac{1}{k}(k^2 \vec{M}_{e1n}) = k\vec{M}_{e1n}$$

we have

$$\begin{aligned} \vec{H}^{inc} &= \frac{jk}{\omega\mu}\sum_{n=0}^{\infty}\left[a_n \vec{N}_{o1n} + b_n \vec{M}_{e1n}\right] \\ &= \frac{j}{\zeta}\sum_{n=0}^{\infty}\left[a_n \vec{N}_{o1n} + b_n \vec{M}_{e1n}\right] \end{aligned} \tag{2.45}$$

Comparing the $\hat{R}$ components of Eqs. (2.44) and (2.45), we have

$$\frac{1}{\zeta}\sin\theta\sin\phi\, e^{-jkR\cos\theta} = \frac{j}{\zeta}\sum_{n=0}^{\infty} a_n \frac{n(n+1)}{kR} J_n(kR) P_n^1(\cos\theta)\sin\phi$$

$$\sin\theta\, e^{-jkR\cos\theta} = j\sum_{n=0}^{\infty} a_n \frac{n(n+1)}{kR} J_n(kR) P_n^1(\cos\theta)$$

$$\because \quad \frac{d}{d\theta}(e^{-jkR\cos\theta}) = jkR\sin\theta\, e^{-jkR\cos\theta}$$

$$\frac{1}{jkR}\frac{d}{d\theta}(e^{-jkR\cos\theta}) = j\sum_{n=0}^{\infty} a_n \frac{n(n+1)}{kR} J_n(kR) P_n^1(\cos\theta)$$

Using Eqs. (2.42) and the relation of $\frac{d}{d\theta}P_n(\cos\theta) = -P_n^1(\cos\theta)$,

$$\sum_{n=1}^{\infty}(-j)^n(2n+1)J_n(kR)P_n^1(\cos\theta) = \sum_{n=1}^{\infty} a_n n(n+1) J_n(kR) P_n^1(\cos\theta)$$

where the terms for $n=0$ are zero in the above equation.

We then have

$$a_n = (-j)^n \frac{2n+1}{n(n+1)} \tag{2.46}$$

Finally, we have

$$\begin{aligned} \vec{E}^{inc} &= \sum_{n=1}^{\infty}\left[a_n \vec{M}_{o1n} + b_n \vec{N}_{e1n}\right] \\ &= \sum_{n=1}^{\infty}(-j)^n \frac{2n+1}{n(n+1)}\left[\vec{M}_{o1n} + j\vec{N}_{e1n}\right] \end{aligned} \tag{2.47}$$

$$\begin{aligned} \vec{H}^{inc} &= \frac{j}{\zeta}\sum_{n=1}^{\infty}\left[b_n \vec{M}_{e1n} + a_n \vec{N}_{o1n}\right] \\ &= \frac{-1}{\zeta}\sum_{n=1}^{\infty}(-j)^n \frac{2n+1}{n(n+1)}\left[\vec{M}_{e1n} - j\vec{N}_{o1n}\right] \end{aligned} \tag{2.48}$$

In Eqs. (2.47) and (2.48), the spherical Bessel function z_n is J_n (the first kind of Bessel function) because $\vec{E}^{inc}$ and $\vec{H}^{inc}$ are finite at R=0. To be more explicit, we can write

$$\vec{E}^{inc} = \sum_{n=1}^{\infty}(-j)^n \frac{2n+1}{n(n+1)}\left[\vec{M}^{(1)}_{o1n} + j\vec{N}^{(1)}_{e1n}\right] \tag{2.49}$$

$$\vec{H}^{inc} = \frac{-1}{\zeta}\sum_{n=1}^{\infty}(-j)^n \frac{2n+1}{n(n+1)}\left[\vec{M}^{(1)}_{e1n} - j\vec{N}^{(1)}_{o1n}\right] \tag{2.50}$$

Now that we have expressed the incident EM field in terms of spherical vector wave functions, we are in a position to determine the scattered EM field by the sphere.

When the incident EM wave illuminates the sphere, there should be a similar series of functions representing the scattered EM wave because the sum of the tangential components of the incident and scattered electric fields must vanish at every point on the surface of the sphere.

It is then reasonable to express the scattered electric field as

$$\vec{E}^S = \sum_{n=1}^{\infty} (-j)^n \frac{2n+1}{n(n+1)} \left[d_n \vec{M}_{o1n} + j e_n \vec{N}_{e1n} \right] \tag{2.51}$$

and

$$\vec{H}^S = \frac{-1}{\zeta} \sum_{n=1}^{\infty} (-j)^n \frac{2n+1}{n(n+1)} \left[e_n \vec{M}_{e1n} - j d_n \vec{N}_{o1n} \right] \tag{2.52}$$

Since the scattered EM fields should behave as

$$e^{-jkR}/R \qquad \text{for large } R\,,$$

the spherical Bessel function in $\vec{M}$ and $\vec{N}$ functions of Eqs. (2.51) and (2.52) should be the second kind of ***Hankel*** function (or the fourth kind of the Bessel function).

$$z_n(kR) = h_n^{(2)}(kR) = \frac{\sqrt{\pi/2}}{\sqrt{kR}} H_{n+\frac{1}{2}}^{(2)}(kR) \rightarrow \frac{e^{-jkR}}{R} \qquad \text{for large } R$$

Note that we designate:

$$z_n^{(1)}(kR) \rightarrow J_n(kR)\,,\; z_n^{(2)}(kR) \rightarrow y_n(kR)$$
$$z_n^{(3)}(kR) \rightarrow h_n^{(1)}(kR)\,,\; z_n^{(4)}(kR) \rightarrow h_n^{(2)}(kR)$$

We can express $\vec{E}^S$ and $\vec{H}^S$ as

$$\vec{E}^S = \sum_{n=1}^{\infty} (-j)^n \frac{2n+1}{n(n+1)} \left[d_n \vec{M}_{o1n}^{(4)} + j e_n \vec{N}_{e1n}^{(4)} \right] \tag{2.53}$$

$$\vec{H}^S = \frac{-1}{\zeta} \sum_{n=1}^{\infty} (-j)^n \frac{2n+1}{n(n+1)} \left[e_n \vec{M}_{e1n}^{(4)} - j d_n \vec{N}_{o1n}^{(4)} \right] \tag{2.54}$$

where $\vec{M}_{o1n}^{(4)}$ implies the $\vec{M}_{o1n}$ function with $h_n^{(2)}(kR)$ as its R-dependence factor.

The boundary conditions on the surface of the sphere, $R=a$, require that the tangential components of the total electric field vanish. That is

$$\left[\vec{E}^{inc}\right]_{\theta}+\left[\vec{E}^{S}\right]_{\theta}=0\text{, on }R=a \tag{2.55}$$

$$\left[\vec{E}^{inc}\right]_{\phi}+\left[\vec{E}^{S}\right]_{\phi}=0\text{, on }R=a \tag{2.56}$$

Eq. (2.55) leads to

$$\begin{aligned}&\sum_{n=1}^{\infty}(-j)^n\frac{2n+1}{n(n+1)}\left\{\left[\frac{1}{\sin\theta}J_n(ka)P_n^1(\cos\theta)\cos\phi\right]\right.\\&\left.\qquad+j\left[\frac{1}{ka}\left[\frac{\partial}{\partial R}(RJ_n(kR))\right]_{R=a}\left[\frac{\partial}{\partial\theta}P_n^1(\cos\theta)\right]\cos\phi\right]\right\}\\&=-\sum_{n=1}^{\infty}(-j)^n\frac{2n+1}{n(n+1)}\left\{d_n\left[\frac{1}{\sin\theta}h_n^{(2)}(ka)P_n^1(\cos\theta)\cos\phi\right]\right.\\&\left.\qquad+je_n\left[\frac{1}{ka}\left[\frac{\partial}{\partial R}\left(Rh_n^{(2)}(kR)\right)\right]_{R=a}\left[\frac{\partial}{\partial\theta}P_n^1(\cos\theta)\right]\cos\phi\right]\right\}\end{aligned} \tag{2.57}$$

Since $P_n^1(\cos\theta)$ functions form a orthogonal set, the values of d_n and e_n should be

$$d_n=-\frac{J_n(ka)}{h_n^{(2)}(ka)} \tag{2.58}$$

$$e_n=-\frac{\left[\frac{\partial}{\partial R}(RJ_n(kR))\right]_{R=a}}{\left[\frac{\partial}{\partial R}\left(Rh_n^{(2)}(kR)\right)\right]_{R=a}} \tag{2.59}$$

On the other hand, Eq. (2.56) leads to

$$\begin{aligned}&\sum_{n=1}^{\infty}(-j)^n\frac{2n+1}{n(n+1)}\left\{\left[-J_n(ka)\left[\frac{\partial}{\partial\theta}P_n^1(\cos\theta)\right]\sin\phi\right]\right.\\&\left.\qquad+j\left[\frac{-1}{ka\sin\theta}\frac{\partial}{\partial R}\left[RJ_n(kR)\right]_{R=a}P_n^1(\cos\theta)\sin\phi\right]\right\}\\&=-\sum_{n=1}^{\infty}(-j)^n\frac{2n+1}{n(n+1)}\left\{d_n\left[-h_n^{(2)}(ka)\left[\frac{\partial}{\partial\theta}P_n^1(\cos\theta)\right]\sin\phi\right]\right.\\&\left.\qquad+je_n\left[\frac{-1}{ka\sin\theta}\frac{\partial}{\partial R}\left[Rh_n^{(2)}(kR)\right]_{R=a}P_n^1(\cos\theta)\sin\phi\right]\right\}\end{aligned} \tag{2.60}$$

Eq. (2.60) will give the same solutions for d_n and e_n as expressed in Eqs. (2.58) and (2.59).

It is now interesting to see what is the R-component of the total electric field on the surface of the sphere.

$$\begin{aligned}&\left[\vec{E}^{inc}\right]_R+\left[\vec{E}^{S}\right]_R\\&=\sum_{n=1}^{\infty}(-j)^n\frac{2n+1}{n(n+1)}\left[j\frac{n(n+1)}{ka}J_n(ka)P_n^1(\cos\theta)\cos\phi\right]\\&+\sum_{n=1}^{\infty}(-j)^n\frac{2n+1}{n(n+1)}\left[je_n\frac{n(n+1)}{ka}h_n^{(2)}(ka)P_n^1(\cos\theta)\cos\phi\right]\neq 0\end{aligned}$$

because e_n as given in Eq. (2.59) will not make corresponding terms of two series cancel.

Therefore, we observe that the tangential components of the total electric field on the surface of the sphere vanish, while the normal (radial) component of the total electric field remains finite, implying a distribution of induced charge on the surface of the sphere.

We may express the final solutions for $\vec{E}^S$ and $\vec{H}^S$ as

$$\begin{aligned}\vec{E}^S&=-\sum_{n=1}^{\infty}(-j)^n\frac{2n+1}{n(n+1)}\left\{\frac{J_n(ka)}{h_n^{(2)}(ka)}\vec{M}_{o1n}^{(4)}+j\frac{\left[\frac{\partial}{\partial R}(RJ_n(kR))\right]_{R=a}}{\left[\frac{\partial}{\partial R}\left(Rh_n^{(2)}(kR)\right)\right]_{R=a}}\vec{N}_{e1n}^{(4)}\right\}\\\vec{H}^S&=\frac{1}{\zeta}\sum_{n=1}^{\infty}(-j)^n\frac{2n+1}{n(n+1)}\left\{\frac{\left[\frac{\partial}{\partial R}(RJ_n(kR))\right]_{R=a}}{\left[\frac{\partial}{\partial R}\left(Rh_n^{(2)}(kR)\right)\right]_{R=a}}\vec{M}_{e1n}^{(4)}-j\frac{J_n(ka)}{h_n^{(2)}(ka)}\vec{N}_{o1n}^{(4)}\right\}\end{aligned}\quad(2.61)$$

Quantities of particular interest are the scattered fields in the far zone of the sphere, i.e. $R \to \infty$.

The asymptotic behaviors of $\vec{M}$ and $\vec{N}$ functions can be found as follows :

Since

$$\lim_{R\to\infty}\left[h_n^{(2)}(kR)\right] \to \frac{j^{n+1}}{kR}e^{-jkR}$$

For large R,

$$\begin{aligned}\vec{M}_{o1n}^{(4)} &\underset{R\to\infty}{\longrightarrow} \frac{1}{\sin\theta}h_n^{(2)}(kR)P_n^1(\cos\theta)\cos\phi\,\hat{\theta} - h_n^{(2)}(kR)\left[\frac{\partial}{\partial\theta}P_n^1(\cos\theta)\right]\sin\phi\,\hat{\phi} \\ &\underset{R\to\infty}{\longrightarrow} \frac{j^{n+1}}{kR}e^{-jkR}\left\{\left[\frac{P_n^1(\cos\theta)}{\sin\theta}\right]\cos\phi\,\hat{\theta} - \left[\frac{\partial}{\partial\theta}P_n^1(\cos\theta)\right]\sin\phi\,\hat{\phi}\right\} \\ \vec{N}_{e1n}^{(4)} &\underset{R\to\infty}{\longrightarrow} \frac{1}{kR}\frac{\partial}{\partial R}\left[Rh_n^{(2)}(kR)\right]\left[\frac{\partial}{\partial\theta}P_n^1(\cos\theta)\right]\cos\phi\,\hat{\theta} \\ &\qquad - \frac{1}{kR\sin\theta}\frac{\partial}{\partial R}\left[Rh_n^{(2)}(kR)\right]P_n^1(\cos\theta)\sin\phi\,\hat{\phi}\end{aligned} \tag{2.62}$$

It is noted that R-component of $\vec{N}_{e1n}^{(4)}$ is ignored, because it behaves as $1/R^2$ in the far zone.

For large R,

$$\frac{\partial}{\partial R}\left[Rh_n^{(2)}(kR)\right] = \frac{\partial}{\partial R}\left[R\frac{j^{n+1}}{kR}e^{-jkR}\right] = \frac{j^{n+1}}{k}(-jk)e^{-jkR} = (-j)j^{n+1}e^{-jkR}$$

So

$$\vec{N}_{e1n}^{(4)} \underset{R\to\infty}{\longrightarrow} \frac{(-j)j^{n+1}}{kR}e^{-jkR}\left\{\left[\frac{\partial}{\partial\theta}P_n^1(\cos\theta)\right]\cos\phi\,\hat{\theta} - \left[\frac{P_n^1(\cos\theta)}{\sin\theta}\right]\sin\phi\,\hat{\phi}\right\} \tag{2.63}$$

Similarly, we can show that

$$\vec{M}^{(4)}_{e1n} \underset{R\to\infty}{\longrightarrow} \frac{j^{n+1}}{kR} e^{-jkR} \left\{ -\left[\frac{P_n^1(\cos\theta)}{\sin\theta}\right] \sin\phi\,\hat{\theta} - \left[\frac{\partial}{\partial\theta} P_n^1(\cos\theta)\right] \cos\phi\,\hat{\phi} \right\} \tag{2.64}$$

$$\vec{N}^{(4)}_{o1n} \underset{R\to\infty}{\longrightarrow} \frac{(-j)j^{n+1}}{kR} e^{-jkR} \left\{ \left[\frac{\partial}{\partial\theta} P_n^1(\cos\theta)\right] \sin\phi\,\hat{\theta} + \left[\frac{P_n^1(\cos\theta)}{\sin\theta}\right] \cos\phi\,\hat{\phi} \right\} \tag{2.65}$$

The backscattered fields in the direction of $\theta=\pi$, or in the direction of the transmitting antenna, is particularly important in the radar technology. In this direction,

$$\frac{P_n^1(\cos\theta)}{\sin\theta} \xrightarrow[\theta=\pi]{} -\frac{(-1)^n}{2} n(n+1)$$

$$\frac{\partial}{\partial\theta} P_n^1(\cos\theta) \xrightarrow[\theta=\pi]{} \frac{(-1)^n}{2} n(n+1)$$

then $\vec{M}$ and $\vec{N}$ functions become

$$\vec{M}^{(4)}_{o1n} \xrightarrow[\substack{R\to\infty\\ \theta=\pi}]{} \frac{-j^{n+1}}{kR} e^{-jkR} \frac{(-1)^n}{2} n(n+1)\left[\cos\phi\,\hat{\theta} + \sin\phi\,\hat{\phi}\right] \tag{2.66}$$

$$\vec{N}^{(4)}_{e1n} \xrightarrow[\substack{R\to\infty\\ \theta=\pi}]{} \frac{(-j)j^{n+1}}{kR} e^{-jkR} \frac{(-1)^n}{2} n(n+1)\left[\cos\phi\,\hat{\theta} + \sin\phi\,\hat{\phi}\right] \tag{2.67}$$

$$\vec{M}^{(4)}_{e1n} \xrightarrow[\substack{R\to\infty\\ \theta=\pi}]{} \frac{j^{n+1}}{kR} e^{-jkR} \frac{(-1)^n}{2} n(n+1)\left[\sin\phi\,\hat{\theta} - \cos\phi\,\hat{\phi}\right] \tag{2.68}$$

$$\vec{N}^{(4)}_{o1n} \xrightarrow[\substack{R\to\infty\\ \theta=\pi}]{} \frac{(-j)j^{n+1}}{kR} e^{-jkR} \frac{(-1)^n}{2} n(n+1)\left[\sin\phi\,\hat{\theta} - \cos\phi\,\hat{\phi}\right] \tag{2.69}$$

Substituting Eqs. (2.66) to (2.69) in Eqs. (2.60) and (2.61), we have

$$\vec{E}^S = \frac{e^{-jkR}}{kR} \sum_{n=1}^{\infty} \frac{j(-1)^n}{2}(2n+1) \left\{ \frac{J_n(ka)}{h_n^{(2)}(ka)} (\cos\phi\,\hat{\theta} + \sin\phi\,\hat{\phi}) - \frac{\left[\frac{\partial}{\partial R}(RJ_n(kR))\right]_{R=a}}{\left[\frac{\partial}{\partial R}\left(Rh_n^{(2)}(kR)\right)\right]_{R=a}} (\cos\phi\,\hat{\theta} + \sin\phi\,\hat{\phi}) \right\}$$

or

$$\vec{E}^S = \frac{je^{-jkR}}{2kR}\cos\phi\sum_{n=1}^{\infty}(-1)^n(2n+1)\left\{\frac{J_n(ka)}{h_n^{(2)}(ka)} - \frac{\left[\frac{\partial}{\partial R}(RJ_n(kR))\right]_{R=a}}{\left[\frac{\partial}{\partial R}\left(Rh_n^{(2)}(kR)\right)\right]_{R=a}}\right\}\hat{\theta}$$
$$+\frac{je^{-jkR}}{2kR}\sin\phi\sum_{n=1}^{\infty}(-1)^n(2n+1)\left\{\frac{J_n(ka)}{h_n^{(2)}(ka)} - \frac{\left[\frac{\partial}{\partial R}(RJ_n(kR))\right]_{R=a}}{\left[\frac{\partial}{\partial R}\left(Rh_n^{(2)}(kR)\right)\right]_{R=a}}\right\}\hat{\phi} \quad (2.70)$$

Fortunately, we can use the ***Wronskian*** of the spherical Bessel functions,

$$J_n(kR)\left[\frac{\partial}{\partial R}\left(Rh_n^{(2)}(kR)\right)\right] - \left[\frac{\partial}{\partial R}(RJ_n(kR))\right]h_n^{(2)}(kR) = \frac{-j}{kR} \quad (2.71)$$

to simplify Eq. (2.70) as follows.

$$\vec{E}^S = \frac{e^{-jkR}}{2kR}\frac{1}{ka}\cos\phi\left\{\sum_{n=1}^{\infty}\frac{(-1)^n(2n+1)}{h_n^{(2)}(ka)\left[\frac{\partial}{\partial R}\left(Rh_n^{(2)}(kR)\right)\right]_{R=a}}\right\}\hat{\theta}$$
$$+\frac{e^{-jkR}}{2kR}\frac{1}{ka}\sin\phi\left\{\sum_{n=1}^{\infty}\frac{(-1)^n(2n+1)}{h_n^{(2)}(ka)\left[\frac{\partial}{\partial R}\left(Rh_n^{(2)}(kR)\right)\right]_{R=a}}\right\}\hat{\phi} \quad (2.72)$$

Similarly,

$$\vec{H}^S = \frac{-e^{-jkR}}{2\zeta kR}\frac{1}{ka}\sin\phi\left\{\sum_{n=1}^{\infty}\frac{(-1)^n(2n+1)}{h_n^{(2)}(ka)\left[\frac{\partial}{\partial R}\left(Rh_n^{(2)}(kR)\right)\right]_{R=a}}\right\}\hat{\theta}$$
$$+\frac{e^{-jkR}}{2\zeta kR}\frac{1}{ka}\cos\phi\left\{\sum_{n=1}^{\infty}\frac{(-1)^n(2n+1)}{h_n^{(2)}(ka)\left[\frac{\partial}{\partial R}\left(Rh_n^{(2)}(kR)\right)\right]_{R=a}}\right\}\hat{\phi} \quad (2.73)$$

Results of Eqs. (2.72) and (2.73) satisfy the following relations :

$$\vec{H}^S = \frac{\hat{R}\times\vec{E}^S}{\zeta}, \text{ or } H_\theta^S = -\frac{E_\phi^S}{\zeta} \text{ and } H_\phi^S = \frac{E_\theta^S}{\zeta}.$$

Since the incident electric field is in the x-direction, we expect the backscattered electric field to be in the x-direction also. In fact, the backscattered electric field in the x-direction can be obtained as

$$E_x^S = \left[E_\theta^S\right]_{\substack{\theta=\pi \\ \phi=\pi}} = -\left[E_\theta^S\right]_{\substack{\theta=\pi \\ \phi=0}}$$
$$= -\left[E_\phi^S\right]_{\substack{\theta=\pi \\ \phi=\pi/2}} = -\left[E_\phi^S\right]_{\substack{\theta=\pi \\ \phi=3\pi/2}}$$

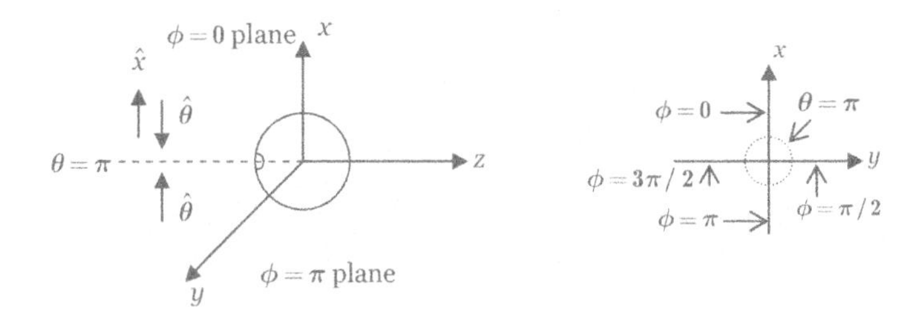

Thus,

$$E_x^S = \left[E_\theta^S\right]_{\substack{\theta=\pi \\ \phi=\pi}} = \frac{-e^{-jkR}}{2kR}\frac{1}{ka}\left\{\sum_{n=1}^{\infty}\frac{(-1)^n(2n+1)}{h_n^{(2)}(ka)\left[\frac{\partial}{\partial R}\left(Rh_n^{(2)}(kR)\right)\right]_{R=a}}\right\} \tag{2.74}$$

In the same plane of $\theta=\pi$ and $\phi=\pi$, the backscattered $\vec{H}$ field can be found to be

$$\left[\vec{H}^S\right]_{\substack{\theta=\pi \\ \phi=\pi}} = \frac{-e^{-jkR}}{2\zeta kR}\frac{1}{ka}\left\{\sum_{n=1}^{\infty}\frac{(-1)^n(2n+1)}{h_n^{(2)}(ka)\left[\frac{\partial}{\partial R}\left(Rh_n^{(2)}(kR)\right)\right]_{R=a}}\right\}\hat{\phi}$$

Since $\left[H_\phi^S\right]_{\substack{\theta=\pi \\ \phi=\pi}} = -H_y^S$,

x
θ = π
y
H_ϕ
φ = π

we have

$$H_y^S = \frac{e^{-jkR}}{2\zeta kR}\frac{1}{ka}\left\{\sum_{n=1}^{\infty}\frac{(-1)^n(2n+1)}{h_n^{(2)}(ka)\left[\frac{\partial}{\partial R}\left(Rh_n^{(2)}(kR)\right)\right]_{R=a}}\right\} \tag{2.75}$$

These results imply that

$\vec{E}^S \times \vec{H}^{S*} \to (-\hat{x}) \times (\hat{y}) \to (-\hat{z}) \to$ propagating in the $(-z)$ direction.

The radar cross section of the sphere is defined as

$$A_e = \lim_{R \to \infty} \left[4\pi R^2\right] \frac{\left|E_x^S\right|^2}{\left|E^{inc}\right|^2} \tag{2.76}$$

From Eq. (2.35),

$$\left|E^{inc}\right|^2 = 1$$

From Eq. (2.74),

$$\left|E_x^S\right|^2 = \left[E_x^S\right]\left[E_x^S\right]^* = \frac{1}{4k^2R^2}\frac{1}{k^2a^2}\left|\sum_{n=1}^{\infty}\frac{(-1)^n(2n+1)}{h_n^{(2)}(ka)\left[\frac{\partial}{\partial R}\left(Rh_n^{(2)}(kR)\right)\right]_{R=a}}\right|^2$$

We can then write A_e as

$$A_e = \frac{\lambda^2}{4\pi(ka)^2}\left|\sum_{n=1}^{\infty}\frac{(-1)^n(2n+1)}{h_n^{(2)}(ka)\left[\frac{\partial}{\partial R}\left(Rh_n^{(2)}(kR)\right)\right]_{R=a}}\right|^2 \tag{2.77}$$

The normalized radar cross section of the sphere, A_e/λ^2, can be plotted as a function of the normalized radius of sphere, a/λ, as in the following figure.

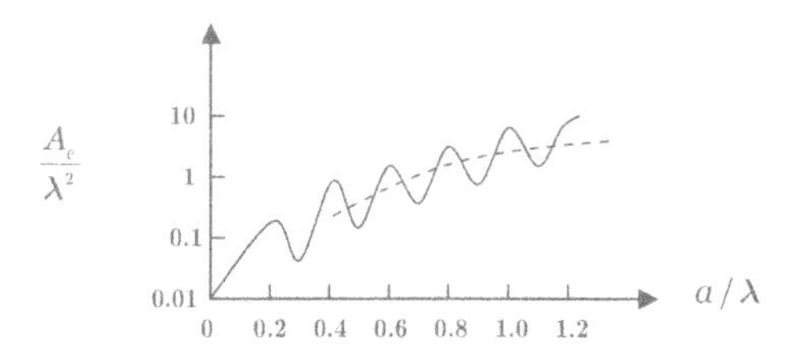

The radar cross section of a small conducting sphere can be obtained easily as follows :

Since

$$h_n^{(2)}(z) = J_n(z) - jy_n(z)$$

as $z \to 0$

$$J_n(z) \to \frac{z^n}{(2n+1)!}$$
$$y_n(z) \to -[(2n-1)!]z^{-(n+1)}$$

So

$$h_n^{(2)}(z) \to -jy_n(z) = j[(2n-1)!]z^{-(n+1)} \quad \because \quad y_n(z) \gg J_n(z)$$

Similarly,

$$\frac{\partial}{\partial z}\left(zh_n^{(2)}(z)\right) \to \frac{\partial}{\partial z}[-jzy_n(z)] = \frac{\partial}{\partial z}\left[j(2n-1)!z^{-n}\right]$$
$$\to -jn(2n-1)!z^{-(n+1)}$$

Thus,

$$h_n^{(2)}(z)\frac{\partial}{\partial z}\left[zh_n^{(2)}(z)\right] \to n[(2n-1)!]^2 z^{-2(n+1)}$$
$$\sum_{n=1}^{\infty}\frac{(-1)^n(2n+1)}{h_n^{(2)}(z)\frac{\partial}{\partial z}\left[zh_n^{(2)}(z)\right]} = \sum_{n=1}^{\infty}\frac{(-1)^n(2n+1)}{n[(2n-1)!]^2}z^{2(n+1)}$$

If $z \to 0$, the infinite series can be well represented by the first term (n=1).

$$\sum_{n=1}^{\infty}\frac{(-1)^n(2n+1)}{n[(2n-1)!]^2}z^{2(n+1)} \to \frac{(-1)(3)}{1[1!]^2}z^{2(2)} = -3z^4$$

Therefore,

$$\lim_{ka\to 0} A_e = \frac{\pi}{k^4a^2}\left|-3(ka)^4\right|^2 = \frac{9\pi}{k^2}(ka)^6 = \frac{9\lambda^2}{4\pi}(ka)^6$$
$$\text{or } = 144\pi^5\left(\frac{a^6}{\lambda^4}\right) \tag{2.78}$$

The result given in Eq. (2.78) is called Rayleigh scattering cross section and this formula is accurate for $a/\lambda < 0.1$. This result implies that the radar cross section of a small conducting sphere varies as λ^{-4}. For a small sphere such as a water vapor, the scattering of light from it has a similar behavior. That is, the higher frequency component of the light scatters much more and that is why the sky is blue.

For a large sphere where $ka >> 1$, the infinite series of Eq. (2.77) converges very poorly. A sophisticated mathematical method is needed to analyze Eq. (2.77) for the case of large ka. This analysis will lead to the findings of the "creeping wave", and the "shadow region" phenomena. Although it is difficult to evaluate the limiting value of A_e for the large ka, it is quite easy to find it based on the physical intuition : A_e should reach the physical optical limit of

$$\lim_{ka\to\infty} A_e = \pi a^2$$

2.e Interaction of a Material Body with EM field in a Rectangular Cavity

As an application of the vector wave functions, we will study the interaction of a material body placed inside an energized rectangular cavity for the purpose of heating the material body.

We will first develop the vector wave functions in rectangular coordinates which describe the rectangular cavity environment. Then, we will derive an electric field integral equation for the induced electric field inside the material body. The electric field integral equation can then be solved numerically.

We will consider a material sample of finite dimensions with dielectric parameters of relative permittivity $\varepsilon = \varepsilon' + j\varepsilon''$, permeability μ, and conductivity σ, and assume that a certain EM mode of a rectangular cavity has been maintained before a material sample is introduced. Our goal is to determine the total EM field inside the material sample induced by the initial cavity EM fields, and the perturbed EM field in the vicinity of the material sample as well.

In many studies involving this type of problems. The unknown induced electric field inside the material sample is expanded in terms of normal cavity electric eigenmodes which are completely solenoidal. This is not correct for the following reason. When a material sample is placed in the cavity, the initial cavity electric field will induce electric changes on the surface of the material sample if it is of finite size or at the heterogeneity boundaries if it is heterogeneous. Thus, the divergence of the electric field will not be zero at the location of the induced charges, or the divergence of electric field will not vanish at all points in the cavity. Therefore, the normal cavity electric eigenmodes which are solenoidal are not sufficient to represent the unknown induced electric field inside the material. Additional eigenmodes which are irrotational will be needed. In this study, a complete set of vector wave functions which include both solenoid and irrotational functions are employed.

In this study, we use the three vector wave functions $\vec{L}_{nml}$, $\vec{M}_{nml}$ and $\vec{N}_{nml}$ as the basis functions to expand the unknown induced electric field inside the cavity.

The definitions of vector wave functions $\vec{L}_{nml}$, $\vec{M}_{nml}$ and $\vec{N}_{nml}$ in rectangular cavities can be found as

$$\vec{L}_{nml} = \frac{1}{k_{nml}}(\nabla \phi^{L}_{nml}) \tag{2.79}$$

$$\vec{M}_{nml} = \nabla \times \hat{z}\phi^{M}_{nml} \tag{2.80}$$

$$\vec{N}_{nml} = \frac{1}{k_{nml}}\nabla \times \nabla \times (\hat{z}\phi^{N}_{nml}) \tag{2.81}$$

where all the scalar wave functions ϕ_{nml} which yield the three vector wave functions satisfy the scalar Helmholtz equation of

$$(\nabla^2 + k^2_{nml})\phi_{nml} = 0 \tag{2.82}$$

and the subscripts n, m and l are used to identify the eigenmodes of a cavity.

The vector wave functions $\vec{L}_{nml}$, $\vec{M}_{nml}$ and $\vec{N}_{nml}$ also need to satisfy the boundary conditions on the perfectly conducting walls of the cavity as :

$$\hat{n} \times \vec{L}_{nml} = 0 \tag{2.83}$$

$$\hat{n} \times \vec{M}_{nml} = 0 \tag{2.84}$$

$$\hat{n} \times \vec{N}_{nml} = 0 \tag{2.85}$$

Based on the definitions of vector wave functions, they have the following properties :

$$\nabla \cdot \vec{M}_{nml} = 0 \tag{2.86}$$

$$\nabla \cdot \vec{N}_{nml} = 0 \tag{2.87}$$

$$\nabla \times \vec{L}_{nml} = 0 \tag{2.88}$$

That is, the vector wave functions $\vec{M}_{nml}$ and $\vec{N}_{nml}$ are solenoidal and $\vec{L}_{nml}$ is irrotational.

The rectangular cavity under consideration has the geometry as shown in Fig.2.

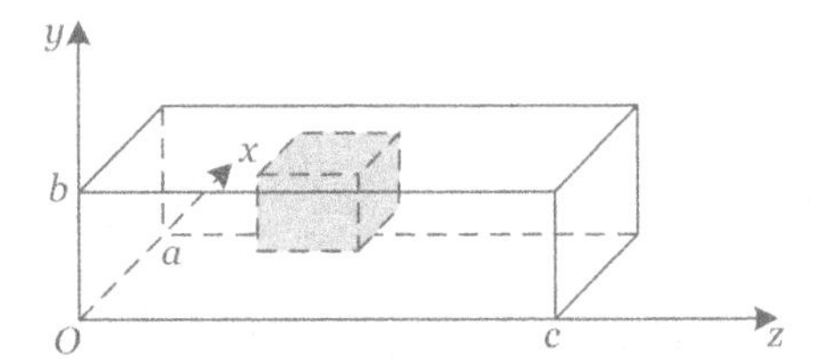

Fig. 2 A material sample inside an energized rectangular cavity.

Before obtaining precise expressions for $\vec{L}_{nml}$, $\vec{M}_{nml}$ and $\vec{N}_{nml}$ in this rectangular cavity environment, we need to prove that these vector wave functions satisfy the vector wave equation. Since the electric field satisfy the vector Helmholtz equation, the basis functions which are used to expand the electric fields should also meet the same requirement. The vector Helmholtz equation is expressed as

$$\nabla^2 \vec{A} + k^2 \vec{A} = 0 \tag{2.89}$$

$$\nabla(\nabla \cdot \vec{A}) - \nabla \times \nabla \times \vec{A} + k^2 \vec{A} = 0 \tag{2.90}$$

We will first prove that $\vec{L}_{nml}$ satisfies the vector Helmholtz equation. Using the definition of $\vec{L}_{nml}$ given in Eq. (2.79),

$$\begin{aligned}\nabla(\nabla \cdot \vec{L}_{nml}) &= \nabla(\nabla \cdot \frac{1}{k_{nml}} \nabla \phi_{nml}) \\ &= \nabla(\frac{1}{k_{nml}} \nabla^2 \phi_{nml}) = -k_{nml} \nabla \phi_{nml} \\ &= -k_{nml}^2 \vec{L}_{nml}\end{aligned}$$

Therefore,

$$\nabla^2 \vec{L}_{nml} + k_{nml}^2 \vec{L}_{nml} = \nabla(\nabla \cdot \vec{L}_{nml}) - \nabla \times \nabla \times \vec{L}_{nml} + k_{nml}^2 \vec{L}_{nml} = 0$$

since $\nabla \times \vec{L}_{nml} \equiv 0$ by definition.

Next, we will prove that $\vec{M}_{nml}$ satisfies the vector Helmholtz equation. Based on the definition of $\vec{M}_{nml}$ given in Eq. (2.80),

$$\vec{M}_{nml} = \nabla \times (\hat{z}\phi^M_{nml}) = \nabla\phi^M_{nml} \times \hat{z} + \phi^M_{nml} \underbrace{\nabla \times \hat{z}}_{0}$$

$$= \hat{x}\frac{\partial}{\partial y}\phi^M_{nml} - \hat{y}\frac{\partial}{\partial x}\phi^M_{nml}$$

$$\nabla^2 \vec{M}_{nml} = \hat{x}\nabla^2(\frac{\partial}{\partial y}\phi^M_{nml}) - \hat{y}\nabla^2(\frac{\partial}{\partial x}\phi^M_{nml})$$

$$= \hat{x}\frac{\partial}{\partial y}(\nabla^2\phi^M_{nml}) - \hat{y}\frac{\partial}{\partial x}(\nabla^2\phi^M_{nml})$$

$$= -k^2_{nml}\hat{x}\frac{\partial}{\partial y}\phi^M_{nml} + k^2_{nml}\hat{y}\frac{\partial}{\partial x}\phi^M_{nml}$$

$$= -k^2_{nml}\vec{M}_{nml}$$

Therefore,

$$\nabla^2 \vec{M}_{nml} + k^2_{nml}\vec{M}_{nml} = 0$$

Lastly, we can show that $\vec{N}_{nml}$ also satisfies the vector Helmholtz equation following the same procedure used in dealing with $\vec{M}_{nml}$.

We will find the expressions for the vector wave functions $\vec{L}_{nml}$, $\vec{M}_{nml}$ and $\vec{N}_{nml}$ in a rectangular cavity.

$$\vec{L}_{nml} = \frac{1}{k_{nml}}(\nabla\phi^L_{nml}) \tag{2.79}$$

$$(\nabla^2 + k^2_{nml})\phi_{nml} = 0 \tag{2.82}$$

Applying the variable separation method to Eq. (2.82), we obtain the solution of the scalar wave function ϕ^L_{nml} as

$$\phi^L_{nml} = A_{nml}\begin{Bmatrix}\cos k_x x\\ \sin k_x x\end{Bmatrix}\begin{Bmatrix}\cos k_y y\\ \sin k_y y\end{Bmatrix}\begin{Bmatrix}\cos k_z z\\ \sin k_z z\end{Bmatrix} \tag{2.89}$$

where A_{nml} is an unknown constant which can be determined by the normalization of the vector wave function $\vec{L}_{nml}$ and

$$k_x^2 + k_y^2 + k_z^2 = k_{nml}^2$$

The three components of the vector wave function $\vec{L}_{nml}$ can be expressed as

$$L_{nmlx} = \frac{A_{nml}}{k_{nml}}k_x\begin{Bmatrix}-\sin k_x x\\ \cos k_x x\end{Bmatrix}\begin{Bmatrix}\cos k_y y\\ \sin k_y y\end{Bmatrix}\begin{Bmatrix}\cos k_z z\\ \sin k_z z\end{Bmatrix} \tag{2.90}$$

$$L_{nmly} = \frac{A_{nml}}{k_{nml}}k_y\begin{Bmatrix}\cos k_x x\\ \sin k_x x\end{Bmatrix}\begin{Bmatrix}-\sin k_y y\\ \cos k_y y\end{Bmatrix}\begin{Bmatrix}\cos k_z z\\ \sin k_z z\end{Bmatrix} \tag{2.91}$$

$$L_{nmlz} = \frac{A_{nml}}{k_{nml}}k_z\begin{Bmatrix}\cos k_x x\\ \sin k_x x\end{Bmatrix}\begin{Bmatrix}\cos k_y y\\ \sin k_y y\end{Bmatrix}\begin{Bmatrix}-\sin k_z z\\ \cos k_z z\end{Bmatrix} \tag{2.92}$$

Based on boundary conditions given by Eq. (2.83), the vector wave function $\vec{L}_{nml}$ can be derived as

$$\begin{aligned}\vec{L}_{nmlz} = \frac{A_{nml}}{k_{nml}}\Big\{&\hat{x}\frac{n\pi}{a}\cos\left(\frac{n\pi}{a}x\right)\sin\left(\frac{m\pi}{b}y\right)\sin\left(\frac{l\pi}{c}z\right)\\ &+\hat{y}\frac{m\pi}{b}\sin\left(\frac{n\pi}{a}x\right)\cos\left(\frac{m\pi}{b}y\right)\sin\left(\frac{l\pi}{c}z\right)\\ &+\hat{z}\frac{l\pi}{c}\sin\left(\frac{n\pi}{a}x\right)\sin\left(\frac{m\pi}{b}y\right)\cos\left(\frac{l\pi}{c}z\right)\Big\}\end{aligned} \tag{2.93}$$

where

$$k_{nml}^2 = \left(\frac{n\pi}{a}\right)^2 + \left(\frac{m\pi}{b}\right)^2 + \left(\frac{l\pi}{c}\right)^2 \tag{2.94}$$

and the expression for the scalar function ϕ_{nml}^{L} is given by

$$\phi_{nml}^{L} = A_{nml} \sin\left(\frac{n\pi}{a}x\right)\sin\left(\frac{m\pi}{b}y\right)\sin\left(\frac{l\pi}{c}z\right) \tag{2.95}$$

The expression for the vector wave function $\vec{M}_{nml}$ in a rectangular cavity can be found from the definition of

$$\vec{M}_{nml} = \nabla \times \hat{z}\phi_{nml}^{M} \tag{2.80}$$

where

$$(\nabla^2 + k_{nml}^2)\phi_{nml}^{M} = 0 \tag{2.82}$$

Using the variable separation method, the solution of the scalar wave function ϕ_{nml}^{M} can be given as

$$\phi_{nml}^{M} = B_{nml} \begin{Bmatrix} \cos k_x x \\ \sin k_x x \end{Bmatrix} \begin{Bmatrix} \cos k_y y \\ \sin k_y y \end{Bmatrix} \begin{Bmatrix} \cos k_z z \\ \sin k_z z \end{Bmatrix} \tag{2.96}$$

where B_{nml} is an unknown constant which can be determined by the normalization of the vector wave function $\vec{M}_{nml}$, and

$$k_x^2 + k_y^2 + k_z^2 = k_{nml}^2$$

The two components of the vector wave function $\vec{M}_{nml}$ can be expressed as

$$M_{nmlx} = B_{nml} k_y \begin{Bmatrix} \cos k_x x \\ \sin k_x x \end{Bmatrix} \begin{Bmatrix} -\sin k_y y \\ \cos k_y y \end{Bmatrix} \begin{Bmatrix} \cos k_z z \\ \sin k_z z \end{Bmatrix} \tag{2.98}$$

$$M_{nmly} = -B_{nml} k_x \begin{Bmatrix} -\sin k_x x \\ \cos k_x x \end{Bmatrix} \begin{Bmatrix} \cos k_y y \\ \sin k_y y \end{Bmatrix} \begin{Bmatrix} \cos k_z z \\ \sin k_z z \end{Bmatrix} \tag{2.99}$$

Based on the boundary conditions given by Eq. (2.84), the vector wave function $\vec{M}_{nml}$ can be derived as

$$\vec{M}_{nml} = B_{nml}\left\{-\hat{x}\frac{m\pi}{b}\cos\left(\frac{n\pi}{a}x\right)\sin\left(\frac{m\pi}{b}y\right)\sin\left(\frac{l\pi}{c}z\right) + \hat{y}\frac{n\pi}{a}\sin\left(\frac{n\pi}{a}x\right)\cos\left(\frac{m\pi}{b}y\right)\sin\left(\frac{l\pi}{c}z\right)\right\} \tag{2.100}$$

where
$$k_{nml}^2 = \left(\frac{n\pi}{a}\right)^2 + \left(\frac{m\pi}{b}\right)^2 + \left(\frac{l\pi}{c}\right)^2 \tag{2.101}$$

and
$$\phi_{nml}^M = B_{nml}\cos\left(\frac{n\pi}{a}x\right)\cos\left(\frac{m\pi}{b}y\right)\sin\left(\frac{l\pi}{c}z\right) \tag{2.102}$$

The expression for the vector wave function $\vec{N}_{nml}$ in a rectangular cavity can be found from the definition of

$$\vec{N}_{nml} = \frac{1}{k_{nml}}\nabla\times\nabla\times\left(\hat{z}\,\phi_{nml}^N\right) \tag{2.81}$$

where
$$(\nabla^2 + k_{nml}^2)\phi_{nml}^N = 0 \tag{2.82}$$

Based on the method of variable separation ϕ_{nml}^N can be obtained as

$$\phi_{nml}^N = C_{nml}\begin{Bmatrix}\cos k_x x\\ \sin k_x x\end{Bmatrix}\begin{Bmatrix}\cos k_y y\\ \sin k_y y\end{Bmatrix}\begin{Bmatrix}\cos k_z z\\ \sin k_z z\end{Bmatrix} \tag{2.103}$$

where C_{nml} is an unknown constant determined by the normalization of $\vec{N}_{nml}$ and

$$k_x^2 + k_y^2 + k_z^2 = k_{nml}^2$$

The three components of the vector wave function $\vec{N}_{nml}$ can be expressed as

$$N_{nmlx} = \frac{C_{nml}}{k_{nml}} k_x k_y \begin{Bmatrix} -\sin k_x x \\ \cos k_x x \end{Bmatrix} \begin{Bmatrix} \cos k_y y \\ \sin k_y y \end{Bmatrix} \begin{Bmatrix} -\sin k_z z \\ \cos k_z z \end{Bmatrix} \tag{2.104}$$

$$N_{nmly} = \frac{C_{nml}}{k_{nml}} k_y k_z \begin{Bmatrix} \cos k_x x \\ \sin k_x x \end{Bmatrix} \begin{Bmatrix} -\sin k_y y \\ \cos k_y y \end{Bmatrix} \begin{Bmatrix} -\sin k_z z \\ \cos k_z z \end{Bmatrix} \tag{2.105}$$

$$N_{nmlz} = \frac{C_{nml}}{k_{nml}} (k_x^2 + k_y^2) \begin{Bmatrix} \cos k_x x \\ \sin k_x x \end{Bmatrix} \begin{Bmatrix} \cos k_y y \\ \sin k_y y \end{Bmatrix} \begin{Bmatrix} \cos k_z z \\ \sin k_z z \end{Bmatrix} \tag{2.106}$$

Based on the boundary condition given by Eq. (2.85), the vector wave function $\vec{N}_{nml}$ can be derived as

$$\begin{aligned} \vec{N}_{nml} = \frac{C_{nml}}{k_{nml}} \Big\{ & -\hat{x} \left(\frac{n\pi}{a}\right)\left(\frac{l\pi}{c}\right) \cos\left(\frac{n\pi}{a}x\right) \sin\left(\frac{m\pi}{b}y\right) \sin\left(\frac{l\pi}{c}z\right) \\ & -\hat{y} \left(\frac{m\pi}{b}\right)\left(\frac{l\pi}{c}\right) \sin\left(\frac{n\pi}{a}x\right) \cos\left(\frac{m\pi}{b}y\right) \sin\left(\frac{l\pi}{c}z\right) \\ & +\hat{z} \left[\left(\frac{n\pi}{a}\right)^2 + \left(\frac{m\pi}{b}\right)^2\right] \sin\left(\frac{n\pi}{a}x\right) \sin\left(\frac{m\pi}{b}y\right) \cos\left(\frac{l\pi}{c}z\right) \Big\} \end{aligned} \tag{2.107}$$

where
$$k_{nml}^2 = \left(\frac{n\pi}{a}\right)^2 + \left(\frac{m\pi}{b}\right)^2 + \left(\frac{l\pi}{c}\right)^2 \tag{2.108}$$

and
$$\phi_{nml}^N = C_{nml} \sin\left(\frac{n\pi}{a}x\right) \sin\left(\frac{m\pi}{b}y\right) \cos\left(\frac{l\pi}{c}z\right) \tag{2.109}$$

From all of these expressions for the vector wave functions, we can identify that $\vec{M}_{nml}$ are the normal TE modes and $\vec{N}_{nml}$ are the normal TM modes in a rectangular cavity. We can also identify $\vec{L}_{nml}$ as the so-called zero frequency modes. It is noted that for these vector eigenfunctions, the eigenvalues

$$k_{nml}^2 = \left(\frac{n\pi}{a}\right)^2 + \left(\frac{m\pi}{b}\right)^2 + \left(\frac{l\pi}{c}\right)^2$$

are the same for the same indices. This will cause some degenerate modes.

Some field structures of the vector wave functions which represent electric fields are given in Ref [1].

The orthogonality and the completeness of the vector wave functions $\vec{L}_{nml}$, $\vec{M}_{nml}$ and $\vec{N}_{nml}$ in a rectangular cavity environment can be proved. These proofs are quite lengthy, and they are also available in Ref [1].

To determine the total electric field induced inside the material sample by the cavity EM field maintained in the rectangular cavity, we need to derive an electric field integral equation for the induced electric field and the associated dyadic Green's function.

The Maxwell's equations in the material sample can be written as

$$\begin{cases} \nabla \times \vec{E}(\vec{r}) = -j\omega\mu_0 \vec{H}(\vec{r}) \\ \nabla \times \vec{H}(\vec{r}) = \sigma\vec{E}(\vec{r}) + j\omega\varepsilon\vec{E}(\vec{r}) \end{cases} \tag{2.110}$$

where $\vec{E}(\vec{r})$ and $\vec{H}(\vec{r})$ are the unknown total electric and magnetic fields in the material sample we aim to determine.

In the empty cavity, the Maxwell's equations are given by

$$\begin{cases} \nabla \times \vec{E}^i(\vec{r}) = -j\omega\mu_0 \vec{H}^i(\vec{r}) \\ \nabla \times \vec{H}^i(\vec{r}) = \ j\omega\varepsilon_0\vec{E}^i(\vec{r}) \end{cases} \tag{2.111}$$

where $\vec{E}^i(\vec{r})$ and $\vec{H}^i(\vec{r})$ are the initial electric and magnetic fields we assumed.

The initial cavity fields will induce electric current and charges inside the material sample. These induced electric currents and charges, in turn, will produce the scattered field or the secondary field $\vec{E}^s(\vec{r})$ and $\vec{H}^s(\vec{r})$. In case, the material sample is of finite size or heterogeneous, there will be induced charges on the sample surface or at the heterogeneity boundaries. Or $\vec{E}^s(\vec{r})$ has an irrotational component.

The total electromagnetic fields $\vec{E}(\vec{r})$ and $\vec{H}(\vec{r})$ can be expressed as

$$\vec{E}(\vec{r}) = \vec{E}^s(\vec{r}) + \vec{E}^i(\vec{r}) \tag{2.112}$$

$$\vec{H}(\vec{r}) = \vec{H}^s(\vec{r}) + \vec{H}^i(\vec{r}) \tag{2.113}$$

Substituting Eqs. (2.112) and (2.113) into Eqs. (2.110) and (2.111) leads to the equations for the scattered fields as

$$\nabla \times \vec{E}^s(\vec{r}) = -j\omega\mu_0 \vec{H}^s(\vec{r}) \tag{2.114}$$

$$\begin{aligned} \nabla \times \vec{H}^s(\vec{r}) &= \sigma\vec{E}(\vec{r}) + j\omega\varepsilon\vec{E}(\vec{r}) - j\omega\varepsilon_0\vec{E}^i(\vec{r}) \\ &= \sigma\vec{E}(\vec{r}) + j\omega(\varepsilon-\varepsilon_0)\vec{E}(\vec{r}) + j\omega\varepsilon_0\vec{E}(\vec{r}) - j\omega\varepsilon_0\vec{E}^i(\vec{r}) \\ &= [\sigma + j\omega(\varepsilon-\varepsilon_0)]\vec{E}(\vec{r}) + j\omega\varepsilon_0[\vec{E}(\vec{r}) - \vec{E}^i(\vec{r})] \\ &= \vec{J}_{eq}(\vec{r}) + j\omega\varepsilon_0\vec{E}^s(\vec{r}) \end{aligned} \tag{2.115}$$

where

$$\vec{J}_{eq}(\vec{r}) \equiv [\sigma + j\omega(\varepsilon-\varepsilon_0)]\vec{E}(\vec{r}) = \tau_e(\vec{r})\vec{E}(\vec{r}) \tag{2.116}$$

$\vec{J}_{eq}(\vec{r})$ is the equivalent current and $\tau_e(\vec{r}) = \sigma + j\omega(\varepsilon-\varepsilon_0)$ is the equivalent complex conductivity.

Taking curl of Eq. (2.114) and using Eq. (2.115), we have

$$\nabla\times\nabla\times\vec{E}^s(\vec{r}) = -j\omega\mu_0\vec{J}_{eq}(\vec{r}) + k_0^2\vec{E}^s(\vec{r}) \tag{2.117}$$

where $k_0^2 = \omega^2\mu_0\varepsilon_0$. Thus, we have the wave equation for the scattered electric field as

$$\nabla\times\nabla\times\vec{E}^s(\vec{r}) - k_0^2\vec{E}^s(\vec{r}) = -j\omega\mu_0\vec{J}_{eq}(\vec{r}) \tag{2.118}$$

We will now proceed to expand $\vec{E}^s(\vec{r})$ in terms of vector wave functions in a rectangular cavity and derive an electric Dyadic Green's function.

The vector wave functions $\vec{L}_{nml}$, $\vec{M}_{nml}$ and $\vec{N}_{nml}$ form a complete set of orthonormal basis functions, satisfy the same boundary conditions as the scattered electric field does and are the solutions of homogeneous vector Helmholtz equation with particular eigenvalues k_{nml}^2. This k_{nml}^2 is not equal to k_0^2 appearing in the inhomogeneous wave equation of Eq. (2.118). However, we can solve Eq. (2.118) by expanding $\vec{E}^s(\vec{r})$ in terms of the vector wave functions $\vec{L}_{nml}$, $\vec{M}_{nml}$ and $\vec{N}_{nml}$. That is

$$\vec{E}^s(\vec{r}) = \sum_n \left[a_n \vec{L}_n(\vec{r}) + b_n \vec{M}_n(\vec{r}) + c_n \vec{N}_n(\vec{r})\right] \tag{2.119}$$

where a_n, b_n and c_n are unknown expansion coefficients. For simplicity, we use one index n instead of three indices n, m and l in the summation of Eq. (2.119).

Substituting Eq. (2.119) into Eq. (2.118) gives

$$\begin{aligned} &\nabla\times\nabla\times\sum_n \left[a_n \vec{L}_n(\vec{r}) + b_n \vec{M}_n(\vec{r}) + c_n \vec{N}_n(\vec{r})\right] \\ &- k_0^2 \sum_n \left[a_n \vec{L}_n(\vec{r}) + b_n \vec{M}_n(\vec{r}) + c_n \vec{N}_n(\vec{r})\right] = -j\omega\mu_0 \vec{J}_{eq}(\vec{r}) \end{aligned} \tag{2.120}$$

Using the properties of the vector wave functions $\vec{L}_{nml}$, $\vec{M}_{nml}$ and $\vec{N}_{nml}$ which we have derived earlier, the above equation can be rewritten as

$$\sum_n \left[-k_0^2 a_n \vec{L}_n(\vec{r}) + b_n (k_n^2 - k_0^2)\vec{M}_n(\vec{r}) + c_n (k_n^2 - k_0^2)\vec{N}_n(\vec{r})\right] = -j\omega\mu_0 \vec{J}_{eq}(\vec{r}) \tag{2.121}$$

Taking the scalar product of Eq. (2.121) with $\vec{L}_{nml}$, $\vec{M}_{nml}$ and $\vec{N}_{nml}$ respectively and integrating over the cavity volume V, then applying the orthonormal properties of the vector wave functions $\vec{L}_{nml}$, $\vec{M}_{nml}$ and $\vec{N}_{nml}$ we obtain the expression for the unknown expansion coefficients as

$$a_n = \frac{j\omega\mu_0}{k_0^2} \int_{V_{sample}} \left[\vec{J}_{eq}(\vec{r}_0) \cdot \vec{L}_n(\vec{r}_0)\right] dV_0 \tag{2.122}$$

$$b_n = \frac{-j\omega\mu_0}{k_n^2 - k_0^2} \int_{V_{sample}} \left[\vec{J}_{eq}(\vec{r}_0) \cdot \vec{M}_n(\vec{r}_0)\right] dV_0 \tag{2.123}$$

$$c_n = \frac{-j\omega\mu_0}{k_n^2 - k_0^2} \int_{V_{sample}} \left[\vec{J}_{eq}(\vec{r}_0) \cdot \vec{N}_n(\vec{r}_0)\right] dV_0 \tag{2.124}$$

where V_{sample} is the volume of the material sample. The integrations of Eqs. (2.122) to (2.124) are carried out over the volume of material sample only because $\vec{J}_{eq}(\vec{r}_0)$ is zero in other points within the cavity volume.

The expression for the scattered electric field $\vec{E}^s(\vec{r})$ can then be expressed as

$$\begin{aligned} \vec{E}^s(\vec{r}) &= -j\omega\mu_0 \int_{V_{sample}} \vec{J}_{eq}(\vec{r}_0) \cdot \sum_n \left[\frac{-\vec{L}_n(\vec{r}_0)\vec{L}_n(\vec{r})}{k_0^2} + \frac{\vec{M}_n(\vec{r}_0)\vec{M}_n(\vec{r}) + \vec{N}_n(\vec{r}_0)\vec{N}_n(\vec{r})}{(k_n^2 - k_0^2)} \right] dV_0 \\ &= -j\omega\mu_0 \int_{V_{sample}} \vec{J}_{eq}(\vec{r}_0) \cdot \overleftrightarrow{G}_e(\vec{r}_0, \vec{r}) dV_0 \end{aligned} \tag{2.125}$$

where the integration region is over the material sample volume.

The electric dyadic Green's function is identified from Eq. (2.125) as

$$\overleftrightarrow{G}_e(\vec{r}_0,\vec{r}) = \sum_n \left[\frac{-\vec{L}_n(\vec{r}_0)\vec{L}_n(\vec{r})}{k_0^2} + \frac{\vec{M}_n(\vec{r}_0)\vec{M}_n(\vec{r}) + \vec{N}_n(\vec{r}_0)\vec{N}_n(\vec{r})}{(k_n^2 - k_0^2)} \right] \tag{2.126}$$

We are now ready to derive an integral equation for the total electric field induced inside the material sample. Based on Eq. (2.116), the expression for the scattered electric field can be expressed as

$$\vec{E}^s(\vec{r}) = -j\omega\mu_0 \int_{V_{sample}} \tau_e(\vec{r}_0)\vec{E}(\vec{r}_0)\cdot \overleftrightarrow{G}_e(\vec{r}_0,\vec{r})dV_0 \tag{2.127}$$

Substituting Eq. (2.127) into Eq. (2.112) gives the electric field integral equation (EFIE) for the unknown total electric field $\vec{E}(\vec{r})$ inside the material sample as

$$\vec{E}(\vec{r}) + j\omega\mu_0 \int_V \tau_e(\vec{r}_0)\vec{E}(\vec{r}_0)\cdot \overleftrightarrow{G}_e(\vec{r}_0,\vec{r})dV_0 = \vec{E}^i(\vec{r}) \tag{2.128}$$

where $\overleftrightarrow{G}_e(\vec{r}_0,\vec{r})$ is given in Eq. (2.126). It is noted that the integration in Eq. (2.128) is over the total cavity volume V. But $\tau_e(\vec{r}_0)$ is not zero only inside the material sample volume V_{sample}, therefore, the integration in Eq. (2.128) is, in effect, over V_{sample}.

The electric field integral equation given in Eq. (2.128) can be modified as follows:

Using $$\sum_n \left[\vec{L}_n(\vec{r}_0)\vec{L}_n(\vec{r}) + \vec{M}_n(\vec{r}_0)\vec{M}_n(\vec{r}) + \vec{N}_n(\vec{r}_0)\vec{N}_n(\vec{r}) \right] = \overleftrightarrow{I}\delta(\vec{r}_0 - \vec{r}) \tag{2.129}$$

which can be found in Ref [1], we can rewrite $\overleftrightarrow{G}_e(\vec{r}_0,\vec{r})$ given in Eq. (2.126) as

$$\begin{aligned} \overleftrightarrow{G}_e(\vec{r}_0,\vec{r}) &= \sum_n \left[k_n^2 \frac{\vec{M}_n(\vec{r}_0)\vec{M}_n(\vec{r}) + \vec{N}_n(\vec{r}_0)\vec{N}_n(\vec{r})}{k_0^2(k_n^2 - k_0^2)} \right] - \frac{\overleftrightarrow{I}\delta(\vec{r}_0 - \vec{r})}{k_0^2} \\ &= \overleftrightarrow{G}_{eo}(\vec{r}_0,\vec{r}) - \frac{\overleftrightarrow{I}\delta(\vec{r}_0 - \vec{r})}{k_0^2} \end{aligned} \tag{2.130}$$

where $$\overleftrightarrow{G}_{eo}(\vec{r}_0,\vec{r}) = \sum_n \left[k_n^2 \frac{\vec{M}_n(\vec{r}_0)\vec{M}_n(\vec{r}) + \vec{N}_n(\vec{r}_0)\vec{N}_n(\vec{r})}{k_0^2(k_n^2 - k_0^2)} \right] \tag{2.131}$$

Therefore, the electric field integral equation (EFIE) of Eq. (2.128) can be rewritten as

$$\vec{E}(\vec{r})\left[1 - \frac{j\omega\mu_0\tau_e(\vec{r})}{k_0^2}\right] + j\omega\mu_0 \int_V \tau_e(\vec{r}_0)\vec{E}(\vec{r}_0)\cdot\overleftrightarrow{G}_{eo}(\vec{r}_0,\vec{r})dV_0 = \vec{E}^i(\vec{r}) \tag{2.132}$$

Eq. (2.132) can be solved numerically once $\tau_e(\vec{r}_0)$ and $\vec{E}^i(\vec{r})$ are given. Some numerical results are available in Ref[1] also.

REFERENCES

[1] J.Zang and K.M.Chen, "Numerical analysis of the induced electric field in a material sample within an energized rectangular cavity," *Journal of EM waves and Application,* vol.13, pp. 1081-1099, 1999

Chapter 3
Dyadic Green's Functions and Applications

In this chapter, we will study the dyadic Green's functions and their applications in solving many problems involving the interaction of EM field with material objects.

3.a Dyadic Analysis

First, we will introduce the dyadic analysis briefly:

A dyadic which is some form of a tensor is defined as

(1)

$$\begin{aligned}
\overleftrightarrow{C} = \vec{A}\vec{B} &= \left(A_x\hat{x} + A_y\hat{y} + A_z\hat{z}\right)\left(B_x\hat{x} + B_y\hat{y} + B_z\hat{z}\right) \\
&= A_xB_x\hat{x}\hat{x} + A_xB_y\hat{x}\hat{y} + A_xB_z\hat{x}\hat{z} \\
&\quad + A_yB_x\hat{y}\hat{x} + A_yB_y\hat{y}\hat{y} + A_yB_z\hat{y}\hat{z} \\
&\quad + A_zB_x\hat{z}\hat{x} + A_zB_y\hat{z}\hat{y} + A_zB_z\hat{z}\hat{z} \\
&= \left(A_x\hat{x} + A_y\hat{y} + A_z\hat{z}\right)B_x\hat{x} \\
&\quad + \left(A_x\hat{x} + A_y\hat{y} + A_z\hat{z}\right)B_y\hat{y} \\
&\quad + \left(A_x\hat{x} + A_y\hat{y} + A_z\hat{z}\right)B_z\hat{z} \\
&= \vec{A}B_x\hat{x} + \vec{A}B_y\hat{y} + \vec{A}B_z\hat{z} \\
&= \vec{A}\sum_{\alpha=1}^{3} B_\alpha\hat{x}_\alpha
\end{aligned}$$

where $\quad \hat{x}_1 = \hat{x},\ \hat{x}_2 = \hat{y},\ \hat{x}_3 = \hat{z}$

We can also write

$$\overleftrightarrow{C} = \vec{C}_1\hat{x} + \vec{C}_2\hat{y} + \vec{C}_3\hat{z}$$

where $\vec{C}_1 = \vec{A}B_x,\ \vec{C}_2 = \vec{A}B_y,\ \vec{C}_3 = \vec{A}B_z$

or

$$\begin{aligned}\overleftrightarrow{C} = {} & C_{xx}\hat{x}\hat{x} + C_{xy}\hat{x}\hat{y} + C_{xz}\hat{x}\hat{z} \\ & + C_{yx}\hat{y}\hat{x} + C_{yy}\hat{y}\hat{y} + C_{yz}\hat{y}\hat{z} \\ & + C_{zx}\hat{z}\hat{x} + C_{zy}\hat{z}\hat{y} + C_{zz}\hat{z}\hat{z}\end{aligned}$$

where $C_{xx} = A_x B_x,\ C_{xy} = A_x B_y,\ C_{xz} = A_x B_z$ etc.

(2)

$$\vec{D}\cdot\overleftrightarrow{C} = \vec{D}\cdot(\vec{A}\vec{B}) = (\vec{D}\cdot\vec{A})\vec{B}$$

$$\overleftrightarrow{C}\cdot\vec{D} = (\vec{A}\vec{B})\cdot\vec{D} = \vec{A}(\vec{B}\cdot\vec{D})$$

So in general $\vec{D}\cdot\overleftrightarrow{C} \neq \overleftrightarrow{C}\cdot\vec{D}$ (unlike $\vec{A}\cdot\vec{B} = \vec{B}\cdot\vec{A}$ in vector)

unless $\overleftrightarrow{C}$ is a symmetric dyadic

i.e. $c_{ij} = c_{ji}$

$$\vec{D}\times\overleftrightarrow{C} = \vec{D}\times(\vec{A}\vec{B}) = (\vec{D}\times\vec{A})\vec{B}$$

$$\overleftrightarrow{C}\times\vec{D} = (\vec{A}\vec{B})\times\vec{D} = \vec{A}(\vec{B}\times\vec{D})$$

(3)

Unit dyadic, $\overleftrightarrow{I}$:

$$\overleftrightarrow{I} = \hat{x}\hat{x} + \hat{y}\hat{y} + \hat{z}\hat{z}$$

$$\vec{A}\cdot\overleftrightarrow{I} = \left(A_x\hat{x} + A_y\hat{y} + A_z\hat{z}\right)\cdot\left(\hat{x}\hat{x} + \hat{y}\hat{y} + \hat{z}\hat{z}\right) = A_x\hat{x} + A_y\hat{y} + A_z\hat{z} = \vec{A}$$

$$\overleftrightarrow{I}\cdot\vec{A} = \vec{A}$$

(4)

$$\begin{aligned}
\nabla\vec{A} &= \nabla\left(A_x\hat{x}+A_y\hat{y}+A_z\hat{z}\right)=\left(\nabla A_x\right)\hat{x}+\left(\nabla A_y\right)\hat{y}+\left(\nabla A_z\right)\hat{z}\\
&=\left(\frac{\partial A_x}{\partial x}\hat{x}+\frac{\partial A_x}{\partial y}\hat{y}+\frac{\partial A_x}{\partial z}\hat{z}\right)\hat{x}\\
&+\left(\frac{\partial A_y}{\partial x}\hat{x}+\frac{\partial A_y}{\partial y}\hat{y}+\frac{\partial A_y}{\partial z}\hat{z}\right)\hat{y}\\
&+\left(\frac{\partial A_z}{\partial x}\hat{x}+\frac{\partial A_z}{\partial y}\hat{y}+\frac{\partial A_z}{\partial z}\hat{z}\right)\hat{z}\\
&= dyadic
\end{aligned}$$

$$\begin{aligned}
\nabla\cdot\overleftrightarrow{C} &= \nabla\cdot\left(\vec{A}\vec{B}\right)=\nabla\cdot\left(\vec{C}_1\hat{x}+\vec{C}_2\hat{y}+\vec{C}_3\hat{z}\right)\\
&=\left(\nabla\cdot\vec{C}_1\right)\hat{x}+\left(\nabla\cdot\vec{C}_2\right)\hat{y}+\left(\nabla\cdot\vec{C}_3\right)\hat{z}\\
&=\left[\nabla\cdot\left(\vec{A}B_x\right)\right]\hat{x}+\left[\nabla\cdot\left(\vec{A}B_y\right)\right]\hat{y}+\left[\nabla\cdot\left(\vec{A}B_z\right)\right]\hat{z}\\
&=\left[\nabla\cdot\vec{A}B_x+\vec{A}\cdot\nabla B_x\right]\hat{x}+\left[\nabla\cdot\vec{A}B_y+\vec{A}\cdot\nabla B_y\right]\hat{y}+\left[\nabla\cdot\vec{A}B_z+\vec{A}\cdot\nabla B_z\right]\hat{z}\\
&=\nabla\cdot\vec{A}\left(B_x\hat{x}+B_y\hat{y}+B_z\hat{z}\right)+\vec{A}\cdot\left(\nabla B_x\hat{x}+\nabla B_y\hat{y}+\nabla B_z\hat{z}\right)\\
&=\left(\nabla\cdot\vec{A}\right)\vec{B}+\vec{A}\cdot\nabla\vec{B}\\
&= vector
\end{aligned}$$

$$\begin{aligned}
\nabla\times\overleftrightarrow{C} &= \nabla\times\left(\vec{A}\vec{B}\right)=\nabla\times\left(\vec{C}_1\hat{x}+\vec{C}_2\hat{y}+\vec{C}_3\hat{z}\right)\\
&=\left(\nabla\times\vec{C}_1\right)\hat{x}+\left(\nabla\times\vec{C}_2\right)\hat{y}+\left(\nabla\times\vec{C}_3\right)\hat{z}\\
&=\left[\nabla\times\left(\vec{A}B_x\right)\right]\hat{x}+\left[\nabla\times\left(\vec{A}B_y\right)\right]\hat{y}+\left[\nabla\times\left(\vec{A}B_z\right)\right]\hat{z}\\
&=\left[\nabla B_x\times\vec{A}+B_x\nabla\times\vec{A}\right]\hat{x}+\left[\nabla B_y\times\vec{A}+B_y\nabla\times\vec{A}\right]\hat{y}+\left[\nabla B_z\times\vec{A}+B_z\nabla\times\vec{A}\right]\hat{z}\\
&=-\vec{A}\times\left[\nabla B_x\hat{x}+\nabla B_y\hat{y}+\nabla B_z\hat{z}\right]+\nabla\times\vec{A}\left(B_x\hat{x}+B_y\hat{y}+B_z\hat{z}\right)\\
&=-\vec{A}\times\nabla\vec{B}+\left(\nabla\times\vec{A}\right)\vec{B}\\
&= dyadic
\end{aligned}$$

(5)

$$
\begin{aligned}
&\int_V \nabla\cdot\overleftrightarrow{C}dV = \oint_s \left(\hat{n}\cdot\overleftrightarrow{C}\right)ds = \oint_s \left(\hat{n}\cdot\vec{A}\vec{B}\right)ds = \oint_s \left(\hat{n}\cdot\vec{A}\right)\vec{B}ds \\
&= \int_V \nabla\cdot\left(\vec{C}_1\hat{x}+\vec{C}_2\hat{y}+\vec{C}_3\hat{z}\right)dV \\
&= \int_V \left[\left(\nabla\cdot\vec{C}_1\right)\hat{x}+\left(\nabla\cdot\vec{C}_2\right)\hat{y}+\left(\nabla\cdot\vec{C}_3\right)\hat{z}\right]dV \\
&= \int_V \left[\left(\nabla\cdot\vec{A}B_x\right)\hat{x}+\left(\nabla\cdot\vec{A}B_y\right)\hat{y}+\left(\nabla\cdot\vec{A}B_z\right)\hat{z}\right]dV \\
&= \hat{x}\int_V \nabla\cdot\left(\vec{A}B_x\right)dV+\hat{y}\int_V \nabla\cdot\left(\vec{A}B_y\right)dV+\hat{z}\int_V \nabla\cdot\left(\vec{A}B_z\right)dV \\
&= \hat{x}\oint_s \hat{n}\cdot\left(\vec{A}B_x\right)ds+\hat{y}\oint_s \hat{n}\cdot\left(\vec{A}B_y\right)ds+\hat{z}\oint_s \hat{n}\cdot\left(\vec{A}B_z\right)ds \\
&= \hat{x}\oint_s \left(\hat{n}\cdot\vec{A}\right)B_x ds+\hat{y}\oint_s \left(\hat{n}\cdot\vec{A}\right)B_y ds+\hat{z}\oint_s \left(\hat{n}\cdot\vec{A}\right)B_z ds \\
&= \oint_s \left(\hat{n}\cdot\vec{A}\right)\left(B_x\hat{x}+B_y\hat{y}+B_z\hat{z}\right)ds \\
&= \oint_s \left(\hat{n}\cdot\vec{A}\right)\vec{B}ds \\
&= vector
\end{aligned}
$$

$$\int_V \nabla\times\overleftrightarrow{C}dV = \oint_s \hat{n}\times\overleftrightarrow{C}ds = \oint_s \hat{n}\times\left(\vec{A}\vec{B}\right)ds = \oint_s \left(\hat{n}\times\vec{A}\right)\vec{B}ds = dyadic$$

$$\int_V \nabla\vec{A}dV = \oint_s \hat{n}\vec{A}ds = dyadic$$

3.b Dyadic Green's Function

The Green's function can be defined as the potential maintained by a point source, such as the scalar potential maintained by a point charge or the vector potential maintained by a current element. These Green's functions are scalar quantities. However, if we want to find the electric field maintained by a current element, we need a dyadic representation because the electric field is not parallel to the current element.

Let's review the scalar Green's function for the scalar potential and vector potential.

The scalar potential $\phi(\vec{r})$ at $\vec{r}$ maintained by a point charge $Q(\vec{r}^{'})$ at $\vec{r}^{'}$.

$$\phi(\vec{r}) = \frac{Q(\vec{r}^{'})e^{-jk_0|\vec{r}-\vec{r}^{'}|}}{4\pi\varepsilon_0|\vec{r}-\vec{r}^{'}|}$$

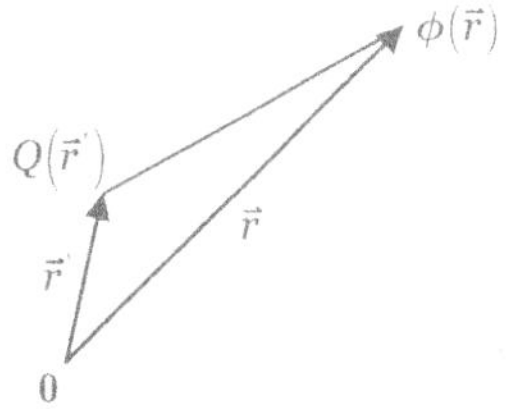

Therefore, the scalar potential maintained by a unit point charge is

$\phi(\vec{r})$
$Q(\vec{r}^{'})$
$\vec{r}$
$\vec{r}^{'}$
0

$$\frac{\phi(\vec{r})}{Q(\vec{r}^{'})} = \frac{e^{-jk_0|\vec{r}-\vec{r}^{'}|}}{4\pi\varepsilon_0|\vec{r}-\vec{r}^{'}|}$$

So we can define

$$G_e(\vec{r}/\vec{r}^{'}) = \frac{e^{-jk_0|\vec{r}-\vec{r}^{'}|}}{4\pi\varepsilon_0|\vec{r}-\vec{r}^{'}|}$$

as the free-space Green's function for scalar potential.

Next, the vector potential $\vec{A}(\vec{r})$ at $\vec{r}$ maintained by a current element with current $\vec{I}$ and length l located at $\vec{r}'$ is given by

$$\vec{A}(\vec{r})=\frac{\mu_0 e^{-jk_0|\vec{r}-\vec{r}'|}}{4\pi|\vec{r}-\vec{r}'|}\vec{I}l$$

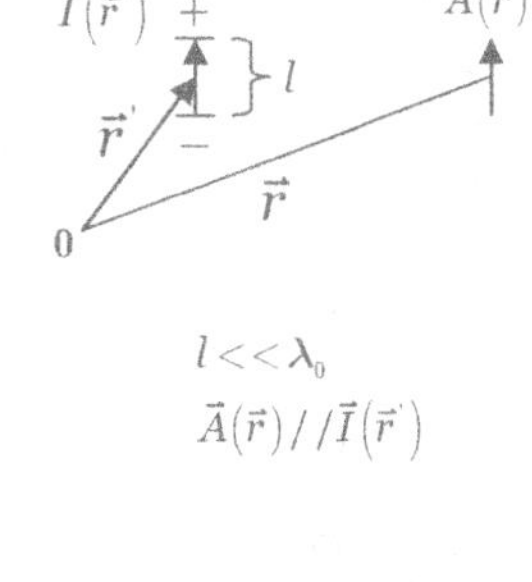

Therefore, the vector potential maintained by a unit current element is

$$\frac{\vec{A}(\vec{r})}{\vec{I}l}=\frac{\mu_0 e^{-jk_0|\vec{r}-\vec{r}'|}}{4\pi|\vec{r}-\vec{r}'|}$$

So we can define

$$G_i(\vec{r}/\vec{r}')=\frac{\mu_0 e^{-jk_0|\vec{r}-\vec{r}'|}}{4\pi|\vec{r}-\vec{r}'|}$$

as the free-space Green's function for the vector potential.

Next, if we want to find the electric field $\vec{E}(\vec{r})$ at $\vec{r}$ maintained by a current element with current $\vec{I}$ and length l located at $\vec{r}'$, it is not a simple operation. First of all, the electric field $\vec{E}(\vec{r})$ and the current element $\vec{I}l$ are not in parallel, second there is no simple relation between $\vec{E}(\vec{r})$ and $\vec{I}l$. That is

$$\frac{\vec{E}(\vec{r})}{\vec{I}l} \neq \text{scalar} = \text{Dyadic}$$

because $\vec{E}(\vec{r})$ is not in parallel to $\vec{I}l$.

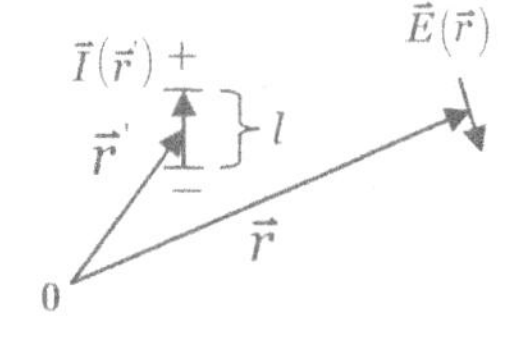

However, we can attempt to find a relation such as

$$\vec{E}(\vec{r})=\overleftrightarrow{G}(\vec{r}/\vec{r}')\cdot\vec{I}l$$

Where $\overleftrightarrow{G}(\vec{r}/\vec{r}')$ is a dyadic function.

or
$$\begin{bmatrix} E_x \\ E_y \\ E_z \end{bmatrix} = \begin{bmatrix} G_{11} & G_{12} & G_{13} \\ G_{21} & G_{22} & G_{23} \\ G_{31} & G_{32} & G_{33} \end{bmatrix} \begin{bmatrix} I_x \\ I_y \\ I_z \end{bmatrix} l$$

or
$$\vec{E}(\vec{r}) = \begin{bmatrix} G_{11}\hat{x}\hat{x} + G_{12}\hat{x}\hat{y} + G_{13}\hat{x}\hat{z} \\ G_{21}\hat{y}\hat{x} + G_{22}\hat{y}\hat{y} + G_{23}\hat{y}\hat{z} \\ G_{31}\hat{z}\hat{x} + G_{32}\hat{z}\hat{y} + G_{33}\hat{z}\hat{z} \end{bmatrix} \cdot \left[I_x\hat{x} + I_y\hat{y} + I_z\hat{z} \right] l$$

Symbolically, the dyadic Green's function is defined as

$$\overleftrightarrow{G} \to \frac{\vec{E}(\vec{r})}{\vec{I}l} \to \text{electric field maintained by unit current element}$$

We then call

$$\overleftrightarrow{G}(\vec{r}/\vec{r}') = \text{free-space dyadic Green's function}$$

Now, how to determine the free-space Green's function?

Let's find $\vec{A}(\vec{r})$ due to $\vec{I}(\vec{r}')l$ first:

$$\vec{A}(\vec{r}) = \frac{\mu_0 e^{-jk_0|\vec{r}-\vec{r}'|}}{4\pi|\vec{r}-\vec{r}'|}\vec{I}l = G(\vec{r}/\vec{r}')\mu_0\vec{I}l$$

where
$$G(\vec{r}/\vec{r}') = \frac{e^{-jk_0|\vec{r}-\vec{r}'|}}{4\pi|\vec{r}-\vec{r}'|} = \textbf{scalar Green's function}$$

$$\begin{aligned}
\vec{E}(\vec{r}) &= -\nabla\phi(\vec{r}) - j\omega\vec{A}(\vec{r}) \\
&= -\nabla\left(\frac{j}{\omega\mu_0\varepsilon_0}\nabla\cdot\vec{A}\right) - j\omega\vec{A} \\
&= -j\omega\vec{A} - \frac{j}{\omega\mu_0\varepsilon_0}\nabla\left(\nabla\cdot\vec{A}\right) \\
&= -j\omega\mu_0 G(\vec{r}/\vec{r}')\vec{I}l - j\frac{1}{\omega\varepsilon_0}\nabla\left[\nabla\cdot\left(G(\vec{r}/\vec{r}')\vec{I}l\right)\right] \\
&= -\omega\mu_0 G(\vec{r}/\vec{r}')\vec{I}l - j\frac{1}{\omega\varepsilon_0}\nabla\left(\nabla G(\vec{r}/\vec{r}')\cdot\vec{I}l\right)
\end{aligned}$$

If $\vec{I}l = \hat{x}I_x l$

$$\begin{aligned}
\vec{E}(\vec{r}) &= -j\omega\mu_0 G(\vec{r}/\vec{r}')I_x l\hat{x} - j\frac{1}{\omega\varepsilon_0}\nabla\left(\frac{\partial}{\partial x}G(\vec{r}/\vec{r}')I_x l\right) \\
&= -j\omega\mu_0 G(\vec{r}/\vec{r}')I_x l\hat{x} - j\frac{1}{\omega\varepsilon_0}\left[\hat{x}\frac{\partial^2 G}{\partial x^2} + \hat{y}\frac{\partial^2 G}{\partial x\partial y} + \hat{z}\frac{\partial^2 G}{\partial x\partial z}\right]I_x l \\
&= \left(-j\omega\mu_0 - j\frac{1}{\omega\varepsilon_0}\frac{\partial^2}{\partial x^2}\right)G(\vec{r}/\vec{r}')I_x l\hat{x} \\
&+ \left(-j\frac{1}{\omega\varepsilon_0}\frac{\partial^2}{\partial x\partial y}\right)G(\vec{r}/\vec{r}')I_x l\hat{y} \\
&+ \left(-j\frac{1}{\omega\varepsilon_0}\frac{\partial^2}{\partial x\partial z}\right)G(\vec{r}/\vec{r}')I_x l\hat{z}
\end{aligned}$$

If $\vec{I}l = \hat{y}I_y l$

$$\begin{aligned}
\vec{E}(\vec{r}) &= -j\omega\mu_0 G(\vec{r}/\vec{r}')I_y l\hat{y} - j\frac{1}{\omega\varepsilon_0}\nabla\left(\frac{\partial}{\partial y}G(\vec{r}/\vec{r}')I_y l\right) \\
&= \left(-j\frac{1}{\omega\varepsilon_0}\frac{\partial^2}{\partial x\partial y}\right)G(\vec{r}/\vec{r}')I_y l\hat{x} \\
&+ \left(-j\omega\mu_0 - j\frac{1}{\omega\varepsilon_0}\frac{\partial^2}{\partial y^2}\right)G(\vec{r}/\vec{r}')I_y l\hat{y} \\
&+ \left(-j\frac{1}{\omega\varepsilon_0}\frac{\partial^2}{\partial y\partial z}\right)G(\vec{r}/\vec{r}')I_y l\hat{z}
\end{aligned}$$

If $\vec{I}l = \hat{z}I_z l$

$$\begin{aligned}
\vec{E}(\vec{r}) &= -j\omega\mu_0 G(\vec{r}/\vec{r}')I_z l\hat{z} - j\frac{1}{\omega\varepsilon_0}\nabla\left(\frac{\partial}{\partial z}G(\vec{r}/\vec{r}')I_z l\right) \\
&= \left(-j\frac{1}{\omega\varepsilon_0}\frac{\partial^2}{\partial x\partial z}\right)G(\vec{r}/\vec{r}')I_z l\hat{x} \\
&+ \left(-j\frac{1}{\omega\varepsilon_0}\frac{\partial^2}{\partial y\partial z}\right)G(\vec{r}/\vec{r}')I_z l\hat{y} \\
&+ \left(-j\omega\mu_0 - j\frac{1}{\omega\varepsilon_0}\frac{\partial^2}{\partial z^2}\right)G(\vec{r}/\vec{r}')I_z l\hat{z}
\end{aligned}$$

Now, if $\vec{I}l = \hat{x}I_x l + \hat{y}I_y l + \hat{z}I_z l$, then

$$\vec{E}(\vec{r}) \to \begin{bmatrix} E_x \\ E_y \\ E_z \end{bmatrix} = \begin{bmatrix} \left(-j\omega\mu_0 - j\frac{1}{\omega\varepsilon_0}\frac{\partial^2}{\partial x^2}\right)G & -j\frac{1}{\omega\varepsilon_0}\frac{\partial^2}{\partial x\partial y}G & -j\frac{1}{\omega\varepsilon_0}\frac{\partial^2}{\partial x\partial z}G \\ -j\frac{1}{\omega\varepsilon_0}\frac{\partial^2}{\partial x\partial y}G & \left(-j\omega\mu_0 - j\frac{1}{\omega\varepsilon_0}\frac{\partial^2}{\partial y^2}\right)G & -j\frac{1}{\omega\varepsilon_0}\frac{\partial^2}{\partial y\partial z}G \\ -j\frac{1}{\omega\varepsilon_0}\frac{\partial^2}{\partial x\partial z}G & -j\frac{1}{\omega\varepsilon_0}\frac{\partial^2}{\partial y\partial z}G & \left(-j\omega\mu_0 - j\frac{1}{\omega\varepsilon_0}\frac{\partial^2}{\partial z^2}\right)G \end{bmatrix} \begin{bmatrix} I_x \\ I_y \\ I_z \end{bmatrix} l$$

If we write $\vec{E}(\vec{r}) = \overleftrightarrow{G}(\vec{r}/\vec{r}')\cdot\vec{I}(\vec{r}')l$

then

$$\overleftrightarrow{G}(\vec{r}/\vec{r}') = \begin{bmatrix} \left(-j\omega\mu_0 - j\frac{1}{\omega\varepsilon_0}\frac{\partial^2}{\partial x^2}\right)G\hat{x}\hat{x} + \left(-j\frac{1}{\omega\varepsilon_0}\frac{\partial^2}{\partial x\partial y}\right)G\hat{x}\hat{y} + \left(-j\frac{1}{\omega\varepsilon_0}\frac{\partial^2}{\partial x\partial z}\right)G\hat{x}\hat{z} \\ \left(-j\frac{1}{\omega\varepsilon_0}\frac{\partial^2}{\partial x\partial y}\right)G\hat{y}\hat{x} + \left(-j\omega\mu_0 - j\frac{1}{\omega\varepsilon_0}\frac{\partial^2}{\partial y^2}\right)G\hat{y}\hat{y} + \left(-j\frac{1}{\omega\varepsilon_0}\frac{\partial^2}{\partial y\partial z}\right)G\hat{y}\hat{z} \\ \left(-j\frac{1}{\omega\varepsilon_0}\frac{\partial^2}{\partial x\partial z}\right)G\hat{z}\hat{x} + \left(-j\frac{1}{\omega\varepsilon_0}\frac{\partial^2}{\partial y\partial z}\right)G\hat{z}\hat{y} + \left(-j\omega\mu_0 - j\frac{1}{\omega\varepsilon_0}\frac{\partial^2}{\partial z^2}\right)G\hat{z}\hat{z} \end{bmatrix}$$

$$= -j\omega\mu_0 \begin{bmatrix} \left(1 + \frac{1}{k_0^2}\frac{\partial^2}{\partial x^2}\right)G\hat{x}\hat{x} + \left(\frac{1}{k_0^2}\frac{\partial^2}{\partial x\partial y}\right)G\hat{x}\hat{y} + \left(\frac{1}{k_0^2}\frac{\partial^2}{\partial x\partial z}\right)G\hat{x}\hat{z} \\ \left(\frac{1}{k_0^2}\frac{\partial^2}{\partial x\partial y}\right)G\hat{y}\hat{x} + \left(1 + \frac{1}{k_0^2}\frac{\partial^2}{\partial y^2}\right)G\hat{y}\hat{y} + \left(\frac{1}{k_0^2}\frac{\partial^2}{\partial y\partial z}\right)G\hat{y}\hat{z} \\ \left(\frac{1}{k_0^2}\frac{\partial^2}{\partial x\partial z}\right)G\hat{z}\hat{x} + \left(\frac{1}{k_0^2}\frac{\partial^2}{\partial y\partial z}\right)G\hat{z}\hat{y} + \left(1 + \frac{1}{k_0^2}\frac{\partial^2}{\partial z^2}\right)G\hat{z}\hat{z} \end{bmatrix}$$

$$= -j\zeta_0 k_0 \left[G\overleftrightarrow{I} + \frac{1}{k_0^2}\nabla\nabla G\right]$$

where $G = G(\vec{r}/\vec{r}') = \frac{e^{-jk_0|\vec{r}-\vec{r}'|}}{4\pi|\vec{r}-\vec{r}'|}$

because $\overleftrightarrow{I} = \hat{x}\hat{x} + \hat{y}\hat{y} + \hat{z}\hat{z}$

$$\begin{aligned}\nabla\nabla G &= \nabla\left(\frac{\partial G}{\partial x}\hat{x}+\frac{\partial G}{\partial y}\hat{y}+\frac{\partial G}{\partial z}\hat{z}\right)\\ &= \left(\nabla\frac{\partial G}{\partial x}\right)\hat{x}+\left(\nabla\frac{\partial G}{\partial y}\right)\hat{y}+\left(\nabla\frac{\partial G}{\partial z}\right)\hat{z}\\ &= \left(\frac{\partial^2 G}{\partial x^2}\hat{x}+\frac{\partial^2 G}{\partial x\partial y}\hat{y}+\frac{\partial^2 G}{\partial x\partial z}\hat{z}\right)\hat{x}\\ &\quad+\left(\frac{\partial^2 G}{\partial x\partial y}\hat{x}+\frac{\partial^2 G}{\partial y^2}\hat{y}+\frac{\partial^2 G}{\partial y\partial z}\hat{z}\right)\hat{y}\\ &\quad+\left(\frac{\partial^2 G}{\partial x\partial z}\hat{x}+\frac{\partial^2 G}{\partial y\partial z}\hat{y}+\frac{\partial^2 G}{\partial z^2}\hat{z}\right)\hat{z}\end{aligned}$$

Now that we have derived

$$\vec{\vec{G}}(\vec{r}/\vec{r}^{'})=-j\zeta_0 k_0\left[\vec{\vec{I}}+\frac{1}{k_0^2}\nabla\nabla\right]G(\vec{r}/\vec{r}^{'})$$

then the electric field due to a current distribution $\vec{J}$ in volume V is

$$\vec{E}(\vec{r})=\int_V \vec{\vec{G}}(\vec{r}/\vec{r}^{'})\cdot\vec{J}(\vec{r}^{'})dV^{'}$$

Notice that $\underset{(\frac{amp}{m^2}m^3)}{\vec{J}dV^{'}} \rightarrow \underset{(amp.m)}{\vec{I}l}$

The dyadic Green's function derived here is valid if the field point $\vec{r}$ is not inside of the source region ($\vec{r}\neq\vec{r}^{'}$). In case $\vec{r}$ is within the source region, $\vec{r}^{'}$ will approach $\vec{r}$ in the integration for $\vec{E}(\vec{r})$. Since the term $\nabla\nabla G(\vec{r}/\vec{r}^{'})$ in $\vec{\vec{G}}(\vec{r}/\vec{r}^{'})$ contains the singularity of order $(\vec{r}-\vec{r}^{'})^{-3}$, the integral will blow up. For this reason, $\vec{\vec{G}}(\vec{r}/\vec{r}^{'})$ inside a source region needs a special attention. It will need the definition of "principal value integration" and a "correction term" to overcome this difficulty.

3.c Dyadic Green's Function in Source Region or Conducting Medium

The dyadic Green's function in a source region or a conducting medium is derived in this section.

Consider a geometry as shown in Fig. 1. A finite volume V of source region or a conducting medium is enclosed by a boundary surface of S. It has electric sources of current density $\vec{J}$ and charge density ρ imbedded inside the volume V, and a surface charge density η residing on the surface S. We aim to determine the electric field $\vec{E}$ at any point inside V.

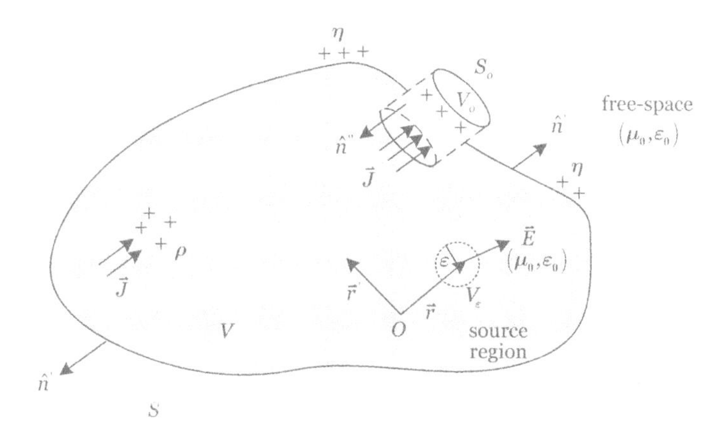

Fig. 1 Geometry of the problem.

$\vec{J}(\vec{r}'), \rho(\vec{r}')=$ given volume densities of electric current and charge in the source region V

$\eta(\vec{r}')=$ surface charge density on the enclosing surface S

$\rho(\vec{r}')$ *and* $\eta(\vec{r}')$ can be expressed in terms of $\vec{J}(\vec{r}')$:

$$\nabla'\cdot\vec{J}(\vec{r}')+j\omega\rho(\vec{r}')=0 \qquad \Rightarrow \qquad \rho(\vec{r}')=\frac{j}{\omega}\nabla'\cdot\vec{J}(\vec{r}') \tag{3.1}$$

Integrating the equation of continuity over a pill box V_o:

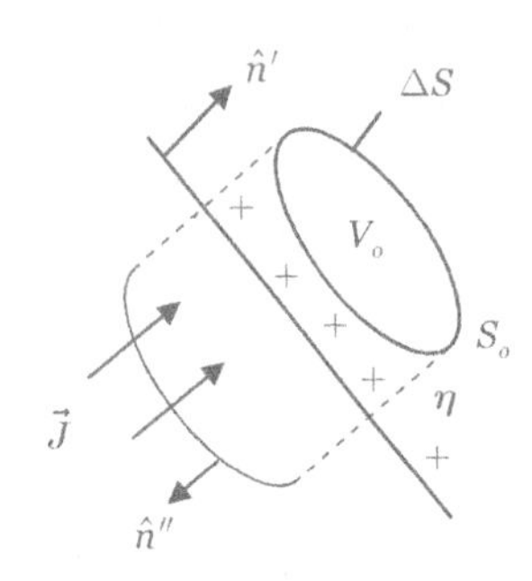

$$\int_{V_o} \nabla' \cdot \vec{J}(\vec{r}\,')dv' = \oint_{S_o} \hat{n}' \cdot \vec{J}(\vec{r}\,')ds' = -j\omega \int_{V_o} \rho(\vec{r}\,')dv'$$

$$\oint_{S_o} \hat{n}' \cdot \vec{J}(\vec{r}\,')ds' = \left[\hat{n}'' \cdot \vec{J}(\vec{r}\,')\right]\Delta s$$

$$= \left[-\hat{n}' \cdot \vec{J}(\vec{r}\,')\right]\Delta s$$

$\because\ \vec{J}(\vec{r}\,') = 0$ on the outer surface

$\hat{n}'' = -\hat{n}'$ on the inner surface

$\int_{V_o} \rho(\vec{r}\,')dv' = \eta(\vec{r}\,')\Delta s$ when the thickness of the pill box goes to zero.

Hence,

$\hat{n}'$ $\eta(\vec{r}\,')$ $\vec{J}(\vec{r}\,')$ S

over the exterior boundary S

$\hat{n}' =$ outgoing unit vector on S

$$\left[-\hat{n}' \cdot \vec{J}(\vec{r}\,')\right]\Delta s = -j\omega\eta(\vec{r}\,')\Delta s$$

or

$$\eta(\vec{r}\,') = \frac{j}{\omega}\left[-\hat{n}' \cdot \vec{J}(\vec{r}\,')\right] \tag{3.2}$$

The electric field at $\vec{r}$, $\vec{E}(\vec{r})$, can be obtained from

$$\vec{E}(\vec{r}) = -\nabla\phi(\vec{r}) - j\omega\vec{A}(\vec{r})$$

$$= -\nabla\left[\int_V \frac{\rho(\vec{r}\,')}{\varepsilon_o} G(\vec{r}/\vec{r}\,')dv' + \int_S \frac{\eta(\vec{r}\,')}{\varepsilon_o} G(\vec{r}/\vec{r}\,')ds'\right] - j\omega \int_V \mu_o \vec{J}(\vec{r}\,')G(\vec{r}/\vec{r}\,')dv' \tag{3.3}$$

where

$$G(\vec{r}/\vec{r}\,') = \frac{e^{-jk_o|\vec{r}-\vec{r}\,'|}}{4\pi|\vec{r}-\vec{r}\,'|}$$

Substitute Eqs. (3.1) and (3.2) in Eq. (3.3),

$$\vec{E}(\vec{r}) = -j\omega\mu_o\left\{\frac{1}{k_o^2}\nabla\left[\int_V \nabla'\cdot\vec{J}(\vec{r}')G(\vec{r}/\vec{r}')dv' - \int_S \hat{n}'\cdot\vec{J}(\vec{r}')G(\vec{r}/\vec{r}')ds'\right] + \int_V \vec{J}(\vec{r}')G(\vec{r}/\vec{r}')dv'\right\}$$

Since

$$\vec{J}(\vec{r}') = \vec{J}(\vec{r}')\cdot\bar{I},\ \ \vec{J}(\vec{r}')G(\vec{r}/\vec{r}') = \vec{J}(\vec{r}')\cdot\bar{I}\ G(\vec{r}/\vec{r}'),$$

thus

$$\vec{E}(\vec{r}) = -j\omega\mu_o\left\{\frac{1}{k_o^2}\left[\int_V \nabla'\cdot\vec{J}(\vec{r}')\nabla G(\vec{r}/\vec{r}')dv' - \int_S \hat{n}'\cdot\vec{J}(\vec{r}')\nabla G(\vec{r}/\vec{r}')ds'\right] + \int_V \vec{J}(\vec{r}')\cdot\bar{I}\ G(\vec{r}/\vec{r}')dv'\right\} \quad (3.4)$$

Since $\vec{r}$ is within the source region V, the integrating point (source point) $\vec{r}'$ will pass through $\vec{r}$ during the integration. When this happens, $|\vec{r}' - \vec{r}| \to 0$ and $G(\vec{r}/\vec{r}')$ and $\nabla G(\vec{r}/\vec{r}')$ both go to infinity. It can be shown that the second volume integral causes no difficulty because its singularity is integrable. However, the first volume integral is not integrable without excluding the singularity point $(\vec{r}' = \vec{r})$ with a small sphere and performing the principal value evaluation. Also the surface integral can be cancelled out in the process of evaluating the first volume integral. Let's consider the first volume integral :

$$\int_V \nabla'\cdot\vec{J}(\vec{r}')\nabla G(\vec{r}/\vec{r}')dv'$$

$$= \lim_{\varepsilon\to 0}\int_{V'=V-V_\varepsilon} \nabla'\cdot\vec{J}(\vec{r}')\nabla G(\vec{r}/\vec{r}')dv' \quad \text{...Principal value integration}$$

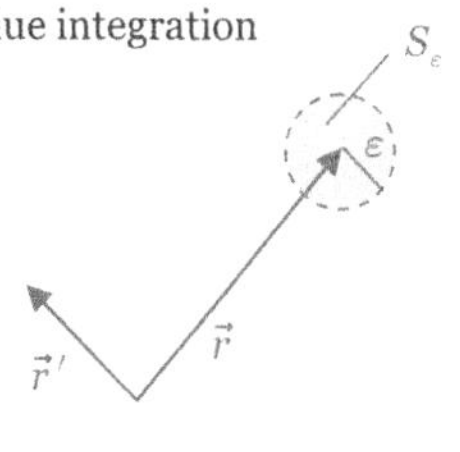

From dyadic identity,

$$\nabla'\cdot(\vec{J}\nabla G) = (\nabla'\cdot\vec{J})\nabla G + \vec{J}\cdot\nabla'\nabla G$$

$$\therefore\ (\nabla'\cdot\vec{J})\nabla G = \nabla'\cdot(\vec{J}\nabla G) - \vec{J}\cdot\nabla'\nabla G$$

Then

$$\lim_{\varepsilon\to 0}\int_{V'=V-V_\varepsilon} \nabla'\cdot\vec{J}(\vec{r}')\nabla G(\vec{r}/\vec{r}')dv'$$

$$=\lim_{\varepsilon\to 0}\left[\int_{V'=V-V_\varepsilon}\nabla'\cdot\left[\vec{J}(\vec{r}')\nabla G(\vec{r}/\vec{r}')\right]dv'-\int_{V'=V-V_\varepsilon}\vec{J}(\vec{r}')\cdot\nabla'\nabla G(\vec{r}/\vec{r}')dv'\right]$$

$$=\lim_{\varepsilon\to 0}\oint_{S+S_\varepsilon}\left[\hat{n}'\cdot\vec{J}(\vec{r}')\right]\nabla G(\vec{r}/\vec{r}')ds'+\lim_{\varepsilon\to 0}\int_{V'=V-V_\varepsilon}\vec{J}(\vec{r}')\cdot\nabla\nabla G(\vec{r}/\vec{r}')dv' \tag{3.5}$$

because

$$\int_V\nabla\cdot\vec{C}dv=\int_V\nabla\cdot(\vec{A}\vec{B})dv=\oint_S(\hat{n}\cdot\vec{A})\vec{B}ds$$

and

$$\nabla'G(\vec{r}/\vec{r}')=-\nabla G(\vec{r}/\vec{r}')$$

Denote,

$$\lim_{\varepsilon\to 0}\int_{V'=V-V_\varepsilon}[\qquad]dv'=P.V.\int_V[\qquad]dv'$$

↑ Principal value

With Eq. (3.5), we can write Eq. (3.4) as

$$\vec{E}(\vec{r})=-j\omega\mu_o\left\{\frac{1}{k_o^2}\left[\int_S\hat{n}'\cdot\vec{J}(\vec{r}')\nabla G(\vec{r}/\vec{r}')ds'+\int_{S_\varepsilon}\hat{n}'\cdot\vec{J}(\vec{r}')\nabla G(\vec{r}/\vec{r}')ds'\right.\right.$$
$$\left.+P.V.\int\vec{J}(\vec{r}')\nabla\nabla G(\vec{r}/\vec{r}')dv'-\int_S\hat{n}'\cdot\vec{J}(\vec{r}')\nabla G(\vec{r}/\vec{r}')ds'\right]$$
$$\left.+P.V.\int_V\vec{J}(\vec{r}')\cdot\vec{I}\,G(\vec{r}/\vec{r}')dv'\right\}$$

Note that

$$\int_V\vec{J}(\vec{r}')\cdot\vec{I}\,G(\vec{r}/\vec{r}')dv'=P.V.\int_V\vec{J}(\vec{r}')\cdot\vec{I}\,G(\vec{r}/\vec{r}')dv'$$

because this integral is not singular

$$\vec{E}(\vec{r}) = -j\zeta_o\kappa_o P.V.\int_V \vec{J}(\vec{r}')\cdot\left[\vec{I} + \frac{\nabla\nabla}{\kappa_o^2}\right]G(\vec{r}/\vec{r}')dv' - \frac{j\zeta_o}{\kappa_o}\lim_{\varepsilon\to 0}\int_{S_\varepsilon} \hat{n}'\cdot\vec{J}(\vec{r}')\nabla G(\vec{r}/\vec{r}')ds' \quad (3.6)$$

Now, we need to evaluate the surface integral on S_ε in Eq. (3.6) carefully. This term will yield a correction term.

Consider

$$\lim_{\varepsilon\to 0}\int_{S_\varepsilon} \hat{n}'\cdot\vec{J}(\vec{r}')\nabla G(\vec{r}/\vec{r}')ds'$$

$$\begin{aligned}\nabla G(\vec{r}/\vec{r}') &= \frac{\partial}{\partial R}\left(\frac{e^{-j\kappa_o R}}{4\pi R}\right)\nabla R \\ &= -\left(\frac{1+j\kappa_o R}{R}\right)\frac{e^{-j\kappa_o R}}{4\pi R}\frac{(\vec{r}-\vec{r}')}{R} \\ &= -\hat{n}'\left(\frac{1+j\kappa_o R}{R}\right)\frac{e^{-j\kappa_o R}}{4\pi R} = -\hat{n}'\frac{1}{4\pi\varepsilon^2}\end{aligned}$$

$\because \; R = \varepsilon \to 0$

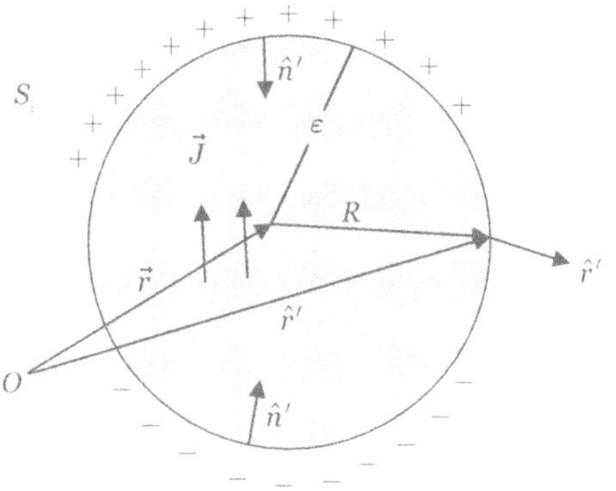

Assuming that $\vec{J}(\vec{r}') = \hat{z}\vec{J}(\vec{r})$ in z-direction and slowing varying

Then

$$\lim_{\varepsilon\to 0}\int_{S_\varepsilon} \hat{n}'\cdot\vec{J}(\vec{r}')\nabla G(\vec{r}/\vec{r}')ds'$$

$$= -\int_{S_\varepsilon} (\hat{n}'\cdot\hat{z})J(\vec{r})\,\hat{n}'\frac{1}{4\pi\varepsilon^2}\varepsilon^2\sin\theta'\,\underbrace{d\theta' d\phi'}_{\theta',\phi' = \text{local spherical coordinates with } \vec{r} \text{ as the center}}$$

$$\begin{aligned}\hat{n}' &= -\hat{r}'(\,\textit{radial}\ \textbf{unit vector on}\ S_\varepsilon) \\ &= -\left[\hat{x}\sin\theta'\cos\phi' + \hat{y}\sin\theta'\sin\phi' + \hat{z}\cos\theta'\right]\end{aligned}$$

$$\hat{n}'\cdot\hat{z} = -\cos\theta'$$

$$\lim_{\varepsilon\to 0}\int_{S_\varepsilon}\hat{n}'\cdot\vec{J}(\vec{r}')\nabla G(\vec{r}/\vec{r}')ds'$$
$$=\frac{J(\vec{r})}{4\pi}\int_0^{2\pi}\int_0^{\pi}-\cos\theta'[\hat{x}\sin\theta'\cos\phi'+\hat{y}\sin\theta'\sin\phi'+\hat{z}\cos\theta']\sin\theta'd\theta'$$
$$=-\hat{z}\frac{J(\vec{r})}{4\pi}\int_0^{2\pi}\int_0^{\pi}\cos^2\theta'\sin\theta'd\theta'd\phi'$$
$$=-\frac{\vec{J}(\vec{r})}{2}\int_0^{\pi}\cos^2\theta'\sin\theta'd\theta' \qquad (\because\ \vec{J}(\vec{r})=\hat{z}J(\vec{r}))$$
$$=-\frac{\vec{J}(\vec{r})}{2}\left[-\frac{\cos^3\theta}{3}\right]_0^{\pi}=-\frac{\vec{J}(\vec{r})}{3} \tag{3.7}$$

Finally, from Eq. (3.6),

$$\vec{E}(\vec{r})=-j\zeta_o\kappa_o P.V.\int_V\vec{J}(\vec{r}')\cdot\left[\overleftrightarrow{I}+\frac{\nabla\nabla}{\kappa_o^2}\right]G(\vec{r}/\vec{r}')dv'+\frac{j\zeta_o}{3\kappa_o}\vec{J}(\vec{r}) \tag{3.8}$$

We have derived the free-space dyadic ***Green's*** function in the previous section as

$$\overleftrightarrow{G}(\vec{r}/\vec{r}')=-j\zeta_o\kappa_o\left[\overleftrightarrow{I}+\frac{\nabla\nabla}{\kappa_o^2}\right]G(\vec{r}/\vec{r}')$$

$$\frac{j\zeta_o}{3\kappa_o}=\frac{j\sqrt{\mu_o/\varepsilon_o}}{3\omega\sqrt{\mu_o\varepsilon_o}}=\frac{j}{3\omega\varepsilon_o}$$

Therefore, we can write

$$\vec{E}(\vec{r})=P.V.\int_V\vec{J}(\vec{r}')\cdot\overleftrightarrow{G}(\vec{r}/\vec{r}')dv'+\frac{j}{3\omega\varepsilon_o}\vec{J}(\vec{r}) \tag{3.9}$$

for $\vec{r}\in V$

where

$$\overleftrightarrow{G}(\vec{r}/\vec{r}')=-j\zeta_o\kappa_o\left[\overleftrightarrow{I}+\frac{\nabla\nabla}{\kappa_o^2}\right]G(\vec{r}/\vec{r}')=\text{ free-space dyadic Green's function}$$

$$G(\vec{r}/\vec{r}')=\frac{e^{-j\kappa_o|\vec{r}'-\vec{r}|}}{|\vec{r}'-\vec{r}|}=\text{ free-space scalar Green's function}$$

The term

$$\frac{j}{3\omega\varepsilon_o}\vec{J}(\vec{r}) = \text{correction term}$$

which is needed when $\vec{r}$ is within the source region.

This term is the current density at the observation point $\vec{r}$ times a constant. It is noted that if $\vec{r}$ is outside the source region, $\vec{J}(\vec{r})=0$ and we don't need a correction term . Furthermore, there is no need for performing a principal value integration because $\overleftrightarrow{G}(\vec{r}/\vec{r}')$ does not go to infinity ($\vec{r}' \neq \vec{r}$). Thus,

$$\vec{E}(\vec{r}) = \int_V \vec{J}(\vec{r}') \cdot \overleftrightarrow{G}(\vec{r}/\vec{r}')dv'$$

$$\text{for } \vec{r} \notin V$$

When the observation point $\vec{r}$ is outside the source region :

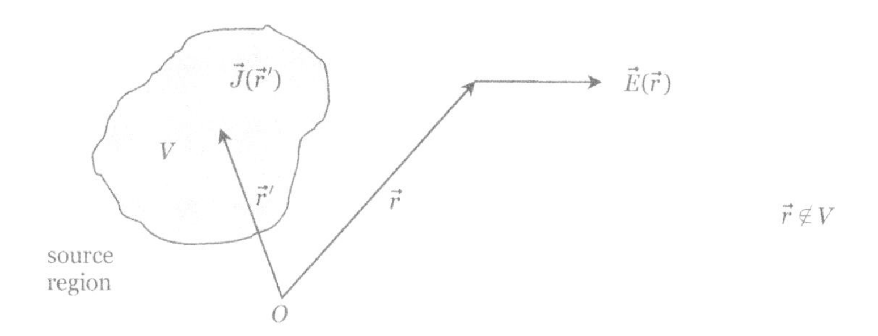

$$\vec{E}(\vec{r}) = \int_V \overleftrightarrow{G}(\vec{r}/\vec{r}') \cdot \vec{J}(\vec{r}')dv'$$

Note :

$$\overleftrightarrow{G}(\vec{r}/\vec{r}') \cdot \vec{J}(\vec{r}') = \vec{J}(\vec{r}') \cdot \overleftrightarrow{G}(\vec{r}/\vec{r}')$$

$$\because\ \overleftrightarrow{G}(\vec{r}/\vec{r}') = symmetric\ dyadic$$

When the observation point $\vec{r}$ is inside the source region :

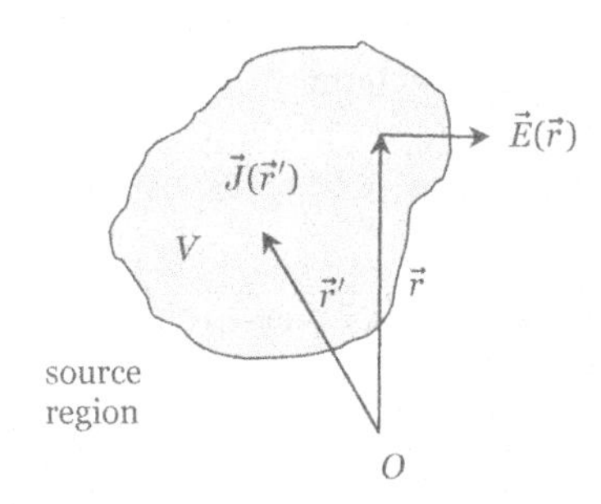

$\vec{r} \in V$

$$\vec{E}(\vec{r}) = P.V.\int_V \vec{G}(\vec{r}/\vec{r}')\cdot\vec{J}(\vec{r}')dv' + \frac{j}{3\omega\varepsilon_o}\vec{J}(\vec{r})$$

3.d Applications of Dyadic Green's Function

Dyadic Green's function has been used in many papers or books published recently. It will gain more popularity in the future. In this section, we will discuss the use of dyadic Green's function in the study of interaction of EM field with a conducting body such as a biological body.

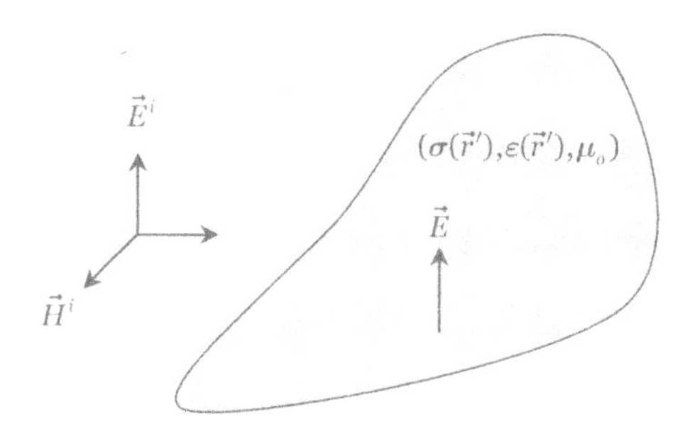

A conducting body is characterized by the parameters $(\sigma(\vec{r}'),\varepsilon(\vec{r}'),\mu_o)$.

The incident electric field $\vec{E}^i$ is incident upon the body and induces a total electric field $\vec{E}$ inside the body.

We aim to find an equation for $\vec{E}$ in terms of $\vec{E}^i$, Maxwell's equation in the body :

$$\begin{cases} \nabla\times\vec{E} = -j\omega\mu_o\vec{H} & (3.10) \\ \nabla\times\vec{H} = \sigma\vec{E} + j\omega\varepsilon\vec{E} & (3.11) \end{cases}$$

Eq. (3.11) can be modified to

$$\nabla\times\vec{H} = \sigma\vec{E} + j\omega(\varepsilon-\varepsilon_o)\vec{E} + j\omega\varepsilon_o\vec{E}$$

$$\begin{aligned}\nabla\times\vec{H} &= [\sigma + j\omega(\varepsilon-\varepsilon_0)]\vec{E} + j\omega\varepsilon_0\vec{E} \\ &= \tau\vec{E} + j\omega\varepsilon_0\vec{E} \\ &= \vec{J}_{eq} + j\omega\varepsilon_0\vec{E}\end{aligned} \tag{3.12}$$

where

$$\begin{aligned}\vec{J}_{eq} &= \tau\vec{E} = [\sigma + \underbrace{j\omega(\varepsilon-\varepsilon_0)}_{\tau}]\vec{E} \\ &= \text{equivalent induced current density}\end{aligned} \tag{3.13}$$

Notice that

$$\sigma\vec{E} = \text{conduction current}$$

$$j\omega(\varepsilon-\varepsilon_0)\vec{E} = \text{polarization current}$$

$$\tau = (\sigma + j\omega(\varepsilon-\varepsilon_0)) = \text{equivalent complex conductivity}$$

From Eqs. (3.10) and (3.12),

$$\nabla\times\vec{E} = -j\omega\mu_0\vec{H} \tag{3.10}$$

$$\nabla\times\vec{H} = \vec{J}_{eq} + j\omega\varepsilon_0\vec{E} \tag{3.12}$$

the body can be considered as a region of equivalent source $\vec{J}_{eq}$ located in the free-space.

Physically, the total electric field $\vec{E}$ is the sum of the incident electric field $\vec{E}^i$ and the scattered (or secondary) electric field $\vec{E}^S$ which is maintained by the induced current $\vec{J}_{eq}$. That is

$$\vec{E}(\vec{r}) = \vec{E}^i(\vec{r}) + \vec{E}^S(\vec{r}) \tag{3.14}$$

Similarly

$$\vec{H}(\vec{r}) = \vec{H}^i(\vec{r}) + \vec{H}^S(\vec{r}) \tag{3.15}$$

The incident EM fields, $\vec{E}^i$ and $\vec{H}^i$, which are the EM fields before the body is introduced, satisfy the following Maxwell equations :

$$\nabla \times \vec{E}^i = -j\omega\mu_0 \vec{H}^i \tag{3.16}$$

$$\nabla \times \vec{H}^i = j\omega\varepsilon_0 \vec{E}^i \tag{3.17}$$

Eq. (3.10) – Eq. (3.16) gives (using Eqs. (3.14) and (3.15))

$$\nabla \times \vec{E}^S = -j\omega\mu_0 \vec{H}^S \tag{3.18}$$

Eq. (3.12) – Eq. (3.17) leads to (using Eqs. (3.15) and (3.14))

$$\nabla \times \vec{H}^S = \vec{J}_{eq} + j\omega\varepsilon_0 \vec{E}^S \tag{3.19}$$

Eqs. (3.18) and (3.19) imply that $\vec{E}^S$ and $\vec{H}^S$ are the scattered EM field maintained by $\vec{J}_{eq}$ inside the source region. This is consistent with the physical intuition as mentioned before. We can then write

$$\vec{E}^S(\vec{r}) = P.V. \int_V \overleftrightarrow{G}(\vec{r} / \vec{r}\,') \cdot \vec{J}_{eq}(\vec{r}\,') dv' + \frac{j}{3\omega\varepsilon_0} \vec{J}_{eq}(\vec{r}) \tag{3.20}$$

Since $\vec{J}_{eq}(\vec{r}) = \tau(\vec{r}\,')\vec{E}(\vec{r}\,')$,

$$\vec{E}^S(\vec{r}) = P.V. \int_V \overleftrightarrow{G}(\vec{r} / \vec{r}\,') \cdot \vec{E}(\vec{r}\,')\tau(\vec{r}\,') dv' + \frac{j\tau(\vec{r})}{3\omega\varepsilon_0} \vec{E}(\vec{r}) \tag{3.21}$$

putting (3.21) in (3.14),

$$\vec{E}(\vec{r}) - \vec{E}^S(\vec{r}) = \vec{E}^i(\vec{r})$$

$$\vec{E}(\vec{r}) - \frac{j\tau(\vec{r})}{3\omega\varepsilon_0}\vec{E}(\vec{r}) - P.V.\int_V \tau(\vec{r}\,')\vec{E}(\vec{r}\,')\cdot\overleftrightarrow{G}(\vec{r}\,/\,\vec{r}\,')dv' = \vec{E}^i(\vec{r})$$

or

$$\left[1 + \frac{\tau(\vec{r})}{j3\omega\varepsilon_0}\right]\vec{E}(\vec{r}) - P.V.\int_V \tau(\vec{r}\,')\vec{E}(\vec{r}\,')\cdot\overleftrightarrow{G}(\vec{r}\,/\,\vec{r}\,')dv' = \vec{E}^i(\vec{r}) \tag{3.22}$$

Eq. (3.22) is called the tensor integral equation for the induced electric field $\vec{E}(\vec{r})$ in a conducting body by an incident electric field $\vec{E}^i(\vec{r})$. This equation was derived by Livesay and Chen [1] in 1974.

Many problems involving the interaction of EM fields with a biological body of arbitrary shape can be solved with Eq. (3.22). For most of the problems, Eq. (3.22) is solved numerically by partitioning the body into many volume cells and assuming $\vec{E}$ field in each cell as an unknown vector quantity. Eq. (3.22) can then be transformed into a set of simultaneous algebraic equations for $\vec{E}$ fields using the Moment Method.

Some papers on this subject are included in references [2] and [3].

3.e Physical Picture of Dyadic Green's Function in Source Region

In the preceding sections, we have shown that a dyadic Green's function can be used to evaluate the electric field in a source region if the principal value of the integral involving the current element is carefully defined and a correction term to the electric field is added.

Let's consider a finite volume of source region V containing a volume current density of $\vec{J}(\vec{r})$ and a volume charge density of $\rho(\vec{r})$ as shown if Fig. 2. The electric field maintained by the source region at an arbitrary point $\vec{r}_1$ outside the source region is given simply as

$$\vec{E}(\vec{r}_1)=\int_V \vec{J}(\vec{r})\cdot\vec{\vec{G}}(\vec{r}_1,\vec{r})dv \tag{3.23}$$

where

$$\vec{\vec{G}}(\vec{r}_1,\vec{r})=-j\omega\mu_0\left[\vec{\vec{I}}+\frac{\nabla\nabla}{k_0^2}\right]\phi(\vec{r}_1,\vec{r})=\text{free space dyadic Green's function}$$

$$\phi(\vec{r}_1,\vec{r})=\frac{e^{-jk_0|\vec{r}_1-\vec{r}|}}{4\pi|\vec{r}_1-\vec{r}|}=\text{free space scalar Green's function}$$

$$k_0=\omega\sqrt{\mu_0\varepsilon_0}$$

If the field point $\vec{r}_0$ is inside the source region, the integral of Eq. (3.23) does not converge. If a small volume surrounding the field point is excluded in the evaluation of the integral, the integral will depend on the shape of the excluded small volume. Usually we can evaluate the electric field at $\vec{r}_0$ as

$$\vec{E}(\vec{r}_0)=PV\int_V \vec{J}(\vec{r})\cdot\vec{\vec{G}}(\vec{r}_0,\vec{r})dv+\vec{E}_C(\vec{r}_0) \tag{3.24}$$

The symbol PV denotes the principal value of the integral meaning the integral is carried out by excluding the small volume surrounding the field point first and then letting the small volume approach to zero. The term $\vec{E}_C(\vec{r}_0)$ is a correction term which should be added to the integral to yield a correct value for the electric field at $\vec{r}_0$. The term $\vec{E}_C(\vec{r}_0)$ is a function of the

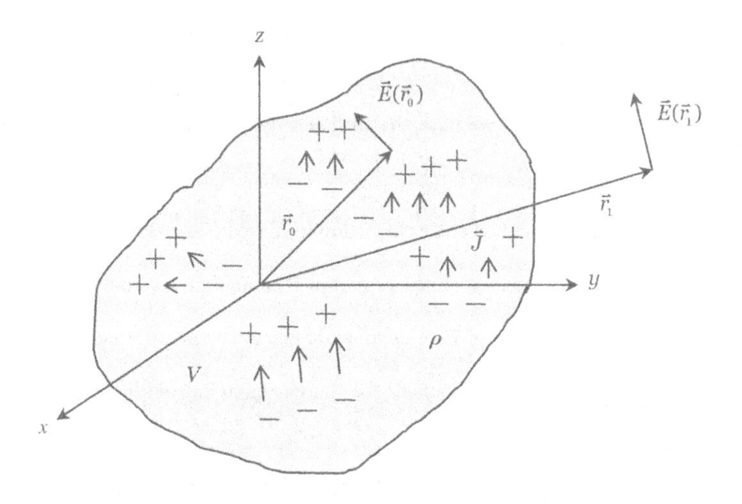

Fig. 2 A source region with volume current density and volume charge density

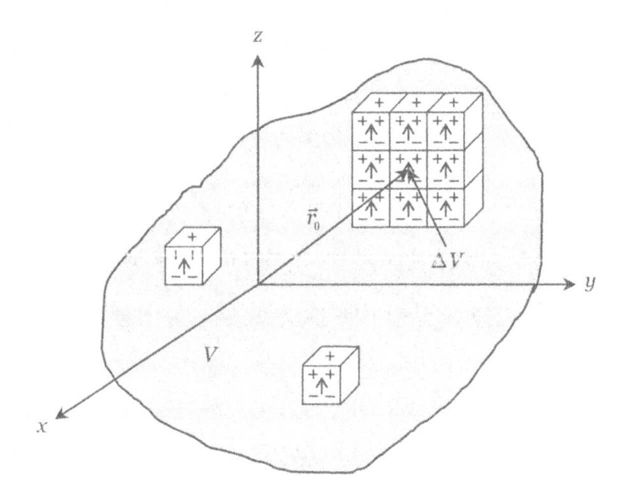

Fig. 3 A source region with a distribution of no uniform current elements

shape of the small volume excluded in the integral evaluation. Some works on this have been previously conducted but they are mainly mathematical in nature and it is difficult to visualize the physical picture involved. For these reasons, we will provide a simple physical picture for a dyadic Green's function in a source region and to define the principal value of the integral and derive the correction term for the electric field based on the some simple physical picture.

The dyadic Green's function is, by definition, the electric field produced by a current element which is actually an electric dipole with a uniform current element and a pair of opposite charges at the ends of the current element. A source region with $\vec{J}$ and ρ such as shown in Fig.2 can be considered as an ensemble of small cells each containing a current element of a electric dipole of certain magnitude and direction as shown in Fig. 3. Based on this picture, two adjoining cells containing current elements of different magnitudes and directions yield a net charge at the interface of these two cells.

Thus a distribution of nonuniform current elements can be made equivalent to a distribution of $\vec{J}$ and ρ.

To determine the electric field at the field point $\vec{r}_0$, a small volume ΔV surrounding $\vec{r}_0$ is excluded to evaluate the principal value of the integral

$$PV\int_V \vec{J}(\vec{r})\cdot\overleftrightarrow{G}(\vec{r}_0,\vec{r})dv \tag{3.25}$$

From the definition of $\overleftrightarrow{G}(\vec{r}_0,\vec{r})$, it is clear that the integral given in Eq. (3.25) is the sum of the electric field maintained at $\vec{r}_0$ by all those current elements in the ensemble of cells except the current element in ΔV. With the integral as defined in Eq. (3.25), the key question is whether the current element in ΔV can maintain an electric field at the center of ΔV or $\vec{r}_0$ as ΔV approaches to zero. The answer is yes, and it is the contribution due to the current element in ΔV which gives rise to the correction term $\vec{E}_C(\vec{r}_0)$.

In the evaluation of the integral of Eq. (3.25), one can choose to use spherical, cylindrical and rectangular coordinate systems depending on the geometry of the source region V and the location of the field point $\vec{r}_0$. It is then, important to choose the shape of ΔV in such a way that the integral of Eq. (3.25) combined with the integration over ΔV will cover the complete source region.

The first case to be considered is a spherical ΔV of radius a as shown in Fig. 4. The correction term $\vec{E}_C(\vec{r}_0)$ is produced by $\vec{J}$ in ΔV and the surface charge η on the spherical surface of ΔV. To evaluate $\vec{E}_C(\vec{r}_0)$, a coordinate system is chosen so that its origin coincide with $\vec{r}_0$ and $\vec{J}$ parallel with the polar axis. Assuming that $\vec{J}$ inside ΔV is uniform, we can write

$$\vec{J} = \hat{z}J \tag{3.26}$$

and the surface charge density on the surface of ΔV can be determined from the equation of continuity as

$$\eta(\theta) = -\frac{j}{\omega} J\cos\theta \tag{3.27}$$

The electric field at the center of ΔV or $\vec{E}_C(\vec{r}_0)$ can be expressed in terms of scalar and vector potential as

$$\vec{E}_C = -\nabla\phi - j\omega\vec{A}$$

If ΔV is very small, the quasistatic approximation can be used to evaluate ϕ and $\vec{A}$ as

$$\phi = \frac{1}{4\pi\varepsilon_0}\int_{\Delta S}\frac{n}{R}ds$$

and

$$\vec{A} = \frac{\mu_0}{4\pi}\int_{\Delta V}\frac{\vec{J}}{R}dv$$

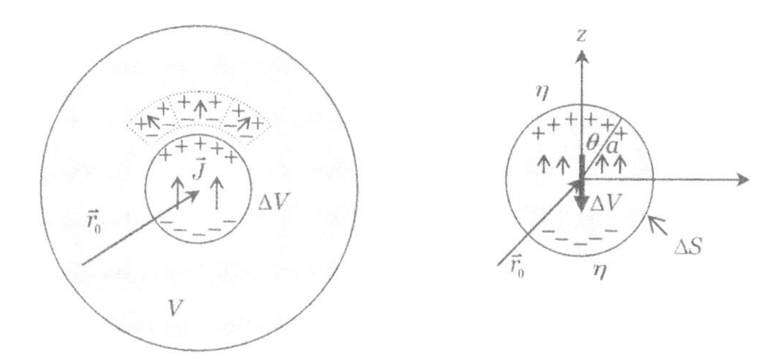

Fig. 4 A spherical ΔV

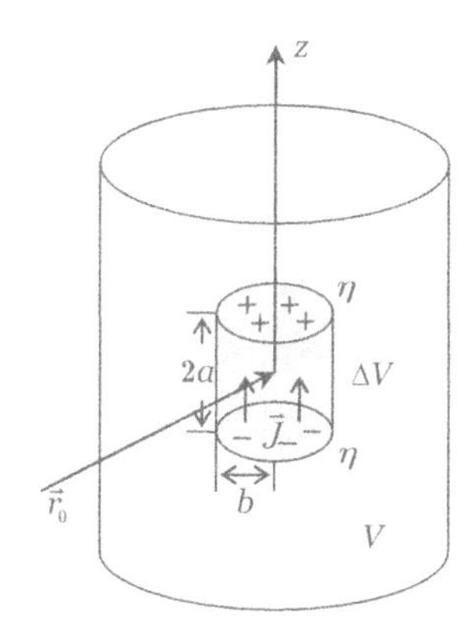

Fig. 5 A cylindrical ΔV

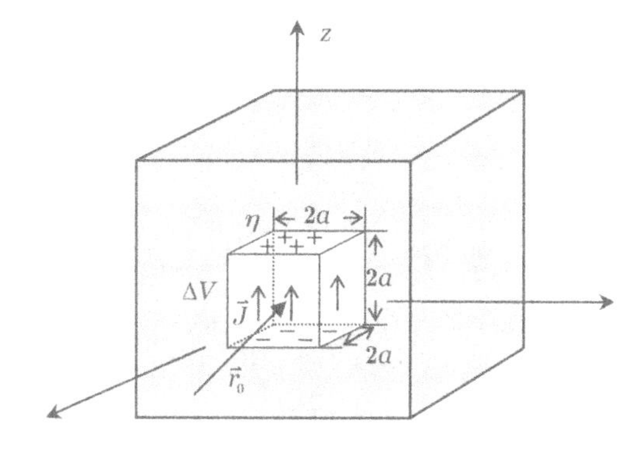

Fig. 6 A rectangular ΔV

where R is the distance between the source point and the field point and ΔS is the surface element of ΔV. It is evident that as ΔV approaches to zero, $\vec{A}$ goes to zero. This implies that the current density $\vec{J}$ does not contribute to the maintenance of $\vec{E}_C(\vec{r}_0)$ in the limit of zero ΔV. The term $-\nabla\phi$ can be easily shown to converge to a finite value as ΔV approach to zero. This means that the surface charge on the surface of ΔV maintains a finite $\vec{E}_C(\vec{r}_0)$ at the center of ΔV as ΔV approaches to zero. For simplicity, we will evaluate the term $-\nabla\phi$ based on the method of finding an electric field. For the case of a spherical ΔV,

$$\begin{aligned}\vec{E}_C(\vec{r}_0) &= \frac{-\hat{z}}{4\pi\varepsilon_0}\int_{\Delta S}\frac{\eta(\theta)}{a^2}\cos\theta\, ds \\ &= \frac{-\hat{z}}{4\pi\varepsilon_0}\left(\frac{-jJ}{\omega}\right)\int_0^{2\pi}\int_0^{\pi}\frac{\cos^2\theta}{a^2}a^2\sin\theta\, d\theta d\phi \\ &= \frac{j\vec{J}(\vec{r}_0)}{3\omega\varepsilon_0}\end{aligned} \tag{3.28}$$

for a spherical ΔV.

The second case is a cylindrical ΔV with a radius b and a height of $2a$ as shown in Fig. 5. $\vec{E}_C(\vec{r}_0)$ in this case is the electric field at the center of ΔV maintained by the surface charges on the top and bottom surfaces of ΔV. If $\vec{J}$ inside of ΔV is given as $\hat{z}\,\vec{J}$, the surface charge can be expressed as

$$\eta = -jJ/\omega$$

Thus,

$$\begin{aligned}\vec{E}_C(\vec{r}_0) &= \lim_{a,b\to 0}\frac{-\hat{z}2}{4\pi\varepsilon_0}\int_0^b\frac{\eta 2\pi r dr}{(a^2+r^2)}\left(\frac{a}{\sqrt{a^2+r^2}}\right) \\ &= \frac{j\vec{J}}{\omega\varepsilon_0}\lim_{a,b\to 0}\left(1-\frac{a}{\sqrt{a^2+b^2}}\right)\end{aligned} \tag{3.29}$$

for a cylindrical ΔV.

If $a = b$,

$$\vec{E}_C(\vec{r}_0) = \frac{j\vec{J}}{\omega\varepsilon_0}(0.293) \tag{3.30}$$

For the one-dimensional geometry, one may make $b >> a$. For this case

$$\vec{E}_C(\vec{r}_0) = \frac{j\vec{J}}{\omega\varepsilon_0} \tag{3.31}$$

The third case is a rectangular ΔV or a small cube with side a as shown in Fig. 6. $\vec{E}_C(\vec{r}_0)$ in this case is the electric field maintained by the surface charges on the upper and lower surfaces of ΔV. With $\vec{J} = \hat{z}J$ and $\eta = -jJ/\omega$, $\vec{E}_C(\vec{r}_0)$ can be determined as

$$\begin{aligned}\vec{E}_C(\vec{r}_0) &= \lim_{a\to 0}\frac{-\hat{z}2}{4\pi\varepsilon_0}\int_{-a}^{a}\int_{-a}^{a}\frac{\eta a\,dx\,dy}{(x^2+y^2+a^2)^{3/2}} \\ &= \frac{j\vec{J}}{\omega\varepsilon_0}\left[\frac{2}{\pi}\int_0^1\frac{ds}{\sqrt{1+s^2}\sqrt{2+s^2}}\right] = \frac{j\vec{J}}{3\omega\varepsilon_0}\end{aligned} \tag{3.32}$$

for a rectangular ΔV. It is noted that $\vec{E}_C(\vec{r}_0)$ for a rectangular ΔV given in Eq. (3.32) is identical to $\vec{E}_C(\vec{r}_0)$ for a spherical ΔV given in Eq. (3.28).

These examples demonstrate a simple method to define the principal value of the integral and evaluate the correction term when a dyadic Green's function is used to compute the electric field in a source region. Using this simple method various coordinate systems can be used and the shape of the excluded ΔV surrounding the field point can be flexible to suit the geometry of the source region.

More information can be found in the paper: Kun-Mu Chen, "a simple picture of Tensor Green's function in Source region", proceeding of the IEEE Aug. 1977. pp1202-1204.

REFERENCES

[1] D. Livesay and K. M. Chen, "Electromagnetic fields induced inside arbitrarily shaped biological bodies," *IEEE Trans. on Microwave Theory and Technique*, vol. MTT-22, no.12, pp. 1273-1280, December 1974.

[2] B. S. Guru and K. M. Chen, "Experimental and theoretical studies on electromagnetic fields induced inside finite biological bodies," *IEEE Trans. on Microwave Theory and Technique*, vol. MTT-24, no.7, pp. 433-440, July 1976.

[3] K. Karimullah, K. M. Chen and D. P. Nyquist, "Electromagnetic coupling between a thin-wire antenna and a neighboring biological bodies," *IEEE Trans. on Microwave Theory and Technique*, vol. MTT-28, no.11, pp. 1218-1225, November 1980.

Chapter 4
Biomedical Application of Electromagnetic Waves

In this chapter we will present an important application of microwave radiation in the biomedical field. Specifically, we will discuss the remote sensing of breathing and heart beats of human subjects located at a distance or behind a barrier. We will study an X-band (10 GHz) life-detection system and a L-band (2 GHz) life-detection system. Also we will discuss a life-detection system operating at (450 MHz and 1.15 GHz) designed especially for locating human subjects buried under earthquake rubble.

Another interesting application of electromagnetic wave theory to quantify the induced current and charge inside a human body exposed to an ELF-LF electric field is also presented in this chapter.

4.a Microwave Life-Detection Systems

The feasibility of the remote sensing of vital signs of human subjects using microwave radiation was demonstrated recently by us at Michigan State University [1], [2]. We will discuss two microwave life-detection systems in this chapter. The first system is an *X*-band (10 GHz) microwave life-detection system which is capable of detecting the breathing and heartbeats of a human subject lying on the ground at a distance of 30 meters or sitting behind a wall of about 6 inches thick. The second system is a *L*-band (2 GHz) microwave life-detection system which was specially designed for detecting the body movements, including the breathing and heartbeats, of human subjects located behind a very thick wall (up to a meter thick). Although these system were originally developed for military and security purposes, they should find some medical applications, especially in the remote physiological sensing area.

The principle on which the systems can be developed is straightforward. We illuminate the subject with a low-intensity (much lower than the safety standard) microwave beam. The small amplitude body movements associated with heartbeat and breathing of the human subject will modulate the backscattered wave, producing a signal from which information of the heart and breathing rates can be extracted using phase detection in the microwave receiving system.

PHYSICAL PRINCIPLES

Some relevant physical principles involved in a physiological sensing system using microwaves are discussed here.

Doppler Effect and Phase Modulation of Microwave Signals

It is well known that when a beam of EM wave is aimed at a moving target, the reflected EM wave form the target will display a frequency shift due to the Doppler effect. The frequency shift is given approximately by

$$\Delta f = f(v/c)$$

where Δf is the frequency shift, f is the frequency of EM wave, v is the velocity of the target relative to the EM wave source, and c is the velocity of light.

The backscattered EM wave can be expressed approximately as

$$E_s = A\cos[2\pi f(1 \pm v/c)t] \tag{4.1}$$

where the positive (negative) sign is used when the target moves toward (away from) the EM source. For example, a police radar gun using a 10 GHz microwave beam can detect a frequency shift of about 1 KHz in the reflected wave from a car traveling at a speed of 60 miles per hour. This frequency shift is sufficiently large to be measured by a conventional frequency detection system such as heterodyne system.

When an EM wave is to be used to detect a very slow movement of a target, such as the body movement associated with heartbeat and breathing, it is impractical to use the Doppler effect, because the frequency shift is extremely small due to an extremely small value of (v/c). For this type of application, it is much more efficient to measure the phase shift in the reflected wave from the slowly moving target. Our microwave life-detection system is based on the detection of the phase modulation in the reflected wave from the human body.

When a microwave beam is incident upon a slowly moving target, the phase angle of the reflected wave will be modulated (or perturbed) bye the target's movement. Mathematically, the reflected wave can be expressed as

$$E_s = A(t)\cos(2\pi ft + \Delta\phi\, u(t)) \tag{4.2}$$

where $\Delta\phi$ is the magnitude of the phase shift and $u(t)$ is a time function which describes the phase variation due to the target's movement. $A(t)$ is the amplitude of the reflected wave and it may also be a function of time due to the target's movement. If a microwave beam is used to detect the body movement due to heartbeat and breathing, the phase shift term ($\Delta\phi u(t)$), is very small compared with the leading term ($2\pi ft$). However, with a phase detection device, such as our microwave life-detection system, this small phase shift can be accurately measured.

A rough relation between the phase shift and the corresponding frequency shift can be given by

$$\Delta f = \frac{\Delta\phi}{2\pi}\frac{\partial u(t)}{\partial t}$$

This relation implies that for body movement due to the heartbeat of 1 Hz, the fundamental frequency shift in the reflected wave of 10 GHz is in the order of 1 Hz, because both $\partial u(t)/\partial t$ and $\Delta\phi/2\pi$ terms are in the order of unity. Obviously, it will be extremely difficult to detect a frequency shift of 1 Hz in a microwave signal of 10 GHz.

Mathematical Formulation of the Phase Modulation of a Reflected EM wave from a Moving Target

To understand how the phase angle of the reflected EM wave is perturbed by the slow movement of the human body, we will analyze the backscattered EM wave from the body when it is illuminated by a plane EM wave. To simplify the problem we model the body as a sphere of complex permittivity. The backscattered field from the sphere is well known [3]. Using the coordinate system shown in Fig.1, the expression for the backscattered electric field may be constructed as

$$\vec{E}_{BS} = \hat{x}\frac{-E_0}{2k_0 r}\sum_{n=1}^{\infty} j^n (2n+1)\cdot\left[-d_n \hat{H}_n^{(2)}(k_0 r) + j e_n \hat{H}_n^{(2)'}(k_0 r)\right] \tag{4.3}$$

where

$$d_n = \frac{\sqrt{\varepsilon_r}\hat{J}_n(k_0 a)\hat{J}'_n(ka) - \hat{J}'_n(k_0 a)\hat{J}_n(ka)}{\hat{H}_n^{(2)'}(k_0 a)\hat{J}_n(ka) - \sqrt{\varepsilon_r}\hat{H}_n^{(2)}(k_0 a)\hat{J}'_n(ka)} \tag{4.4}$$

$$e_n = \frac{\hat{J}_n(k_0 a)\hat{J}'_n(ka) - \sqrt{\varepsilon_r}\hat{J}'_n(k_0 a)\hat{J}_n(ka)}{\sqrt{\varepsilon_r}\hat{H}_n^{(2)'}(k_0 a)\hat{J}_n(ka) - \hat{H}_n^{(2)}(k_0 a)\hat{J}'_n(ka)} \tag{4.5}$$

$$\hat{J}_n(x) = \sqrt{\frac{\pi x}{2}} J_{n+1/2}(x) \tag{4.6}$$

$$\hat{H}_n^{(2)}(x) = \sqrt{\frac{\pi x}{2}} H_{n+1/2}^{(2)}(x) \tag{4.7}$$

ε_r is the complex permittivity of the sphere, a is its radius, k and k_0 represent wavenumbers inside and outside the sphere, respectively, and usual notations for Bessel functions and their derivatives are employed. E_0 is the amplitude of the incident plane wave and $\hat{x}$ is the unit vector along the x axis.

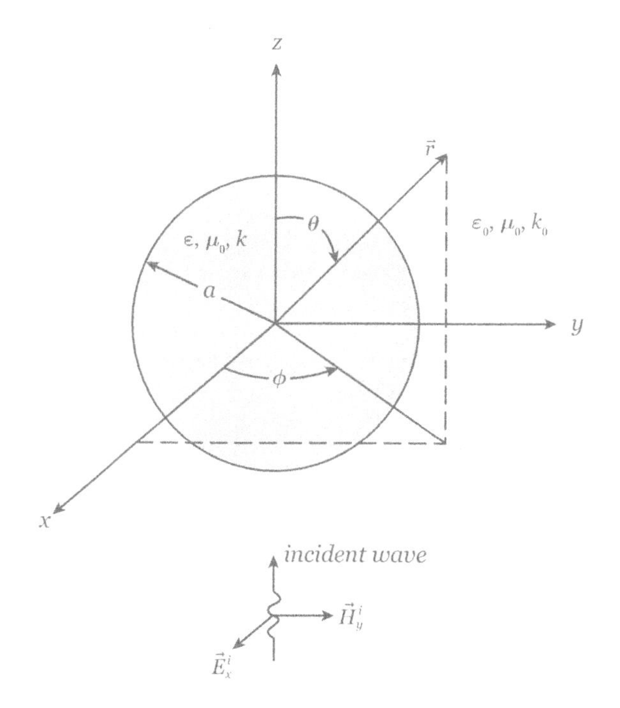

Fig. 1 A conducting sphere illuminated by an incident EM wave.

The phase and the square of the magnitude of the backscattered electric field $\vec{E}_{BS}$ from a sphere of relative permittivity 39.9 and conductivity 10.3 S/m are depicted in Fig. 2 as functions of the radius multiplied by the wavenumber k_0 of the medium. The frequency of the microwave radiation is assumed to be 10 GHz, and the sphere is situated 30.48 m (100 ft) from the transceiver. Breathing and heartbeat produce small vibrations of the spherical surface due to changes in its radius. From Fig. 2, we conclude that these vibrations will produce a linear change in the phase and a relatively smaller linear change in the amplitude squared of the backscattered field. Similar results were obtained when the body was modeled as an infinitely long cylinder of complex permittivity, illuminated by a TM-polarized plane electromagnetic wave.

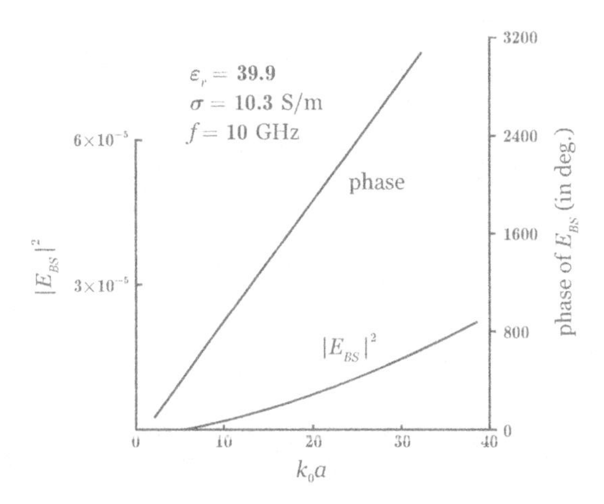

Fig. 2 Phase and magnitude squared of the backscattered field E_{BS} from a sphere as a function of k_0a at 10 GHz at a distance of 30.48 m.

These simplified models show that there will be, in general, amplitude as well as phase modulation of the incident wave as it is backscattered by the body. However, since the phase variation is more linear and it is easier to detect the phase variation from the viewpoint of the signal/noise ratio, we used the phase modulation of the backscattered wave to find the vibrations of the body surface caused by the heartbeat and breathing.

AN X-BAND MICORWAVE LIFE-DETECTION SYSTEM

In this section, the circuit diagram and operation principle of an X-band (GHz) microwave life-detection system will be described. Typical results on the measurement of heart and breathing signals will also be given.

Circuit Diagram and Operation Principle

The schematic diagram of the X-band life-detection system is shown in Fig. 3. A phase-locked oscillator at 10 GHz produces a stable output of about 20 mW. This output is amplified by a low-noise microwave amplifier to a power level of about 200 mW. The output of the amplifier is fed through a 6 dB directional coupler, a variable attenuator, a circulator, and then to a horn antenna. The 6 dB directional coupler branches out 1/4 of the amplifier output to provide a reference signal for clutter cancellation and another reference signal for the mixer. The variable attenuator controls the power level of the microwave signal to be radiated by the antenna. Usually, the radiated power is kept at a level of about 10-20 mW. The horn antenna radiates a microwave beam of about 15° beam-width aimed at the human subjects to be monitored.

The signal received by the antenna consists of a large clutter and a weak return signal scattered from the body. To be able to detect the weak signal modulated by the body movement, the large background clutter needs to be cancelled. This is accomplished by an automatic clutter cancellation circuit which consists of a variable phase-shifter, a variable attenuator and a microprocessor unit which digitally controls the former two components. This automatic clutter cancellation circuit provides an optimal reference signal which is mixed with the received signal by the antenna in a 10 dB directional coupler for the purpose of canceling the clutter. The output of the 10 dB directional coupler contains mainly the weak scattered signal from the body. This body scattered signal is a 10 GHz CW microwave signal modulated by the breathing and the heartbeat. This signal is then amplified by a low noise microwave preamplifier (30 dB) and then mixed with another reference signal in a double-balanced mixer.

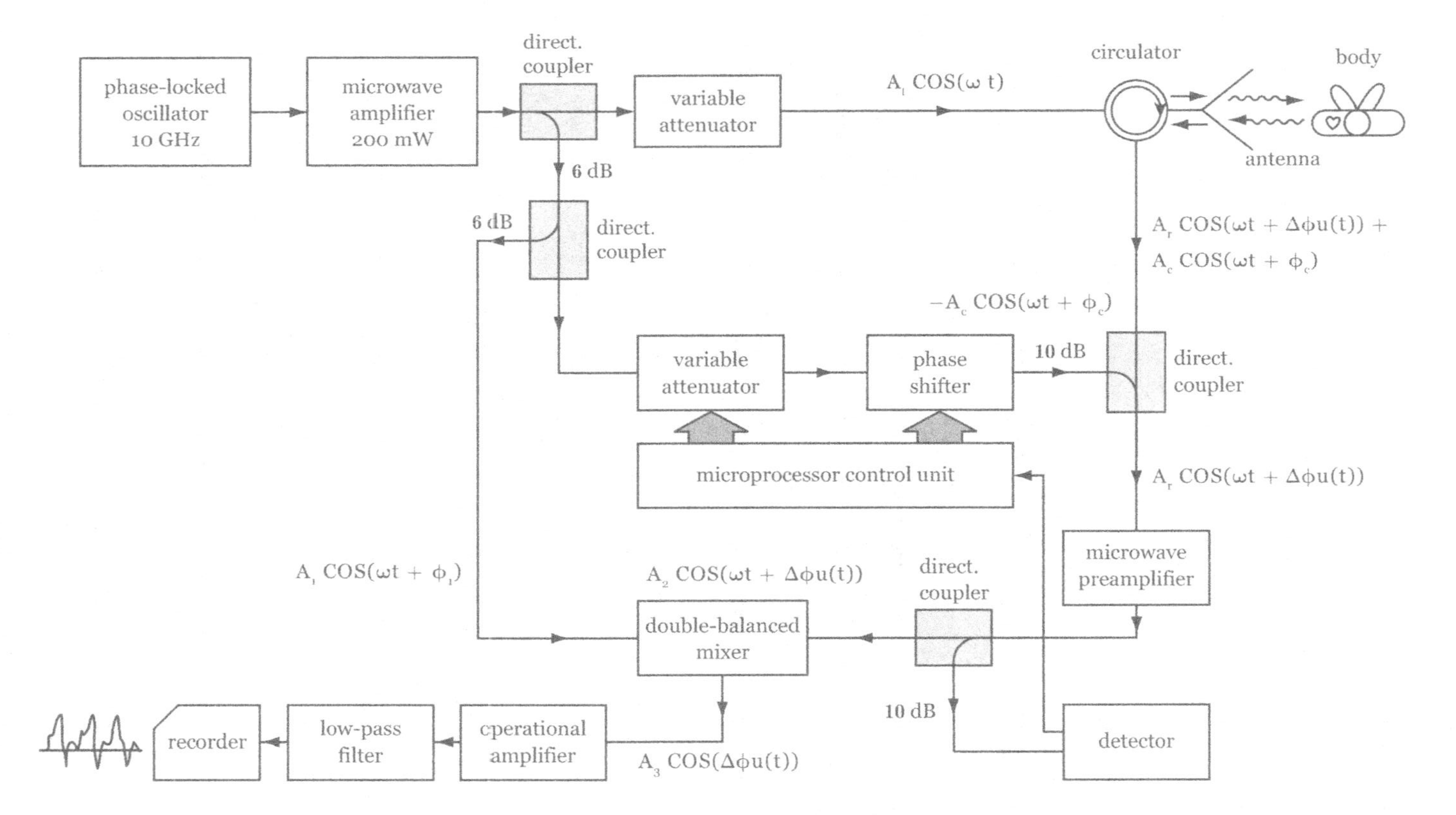

Fig. 3 Circuit diagram of the microwave life-detection system.

Reprinted from "An X-band microwave life-detection system" by Kun-Mu Chen, D. Misra, H. Wang, H. R. Chuang and E. Postow, IEEE Trans. on Biomedical Engineer-ing, Vol. BME-33, No. 7, pp 697-701, July 1986. (©1986 IEEE.)

Between the microwave preamplifier and the double-balanced mixer, a 10 dB directional coupler is inserted to take out a small portion of the amplifier signal for providing an input to the microprocessor unit which controls the phase-shifter and the attenuator. The optimal settings for the phase-shifter and the attenuator are determined by this input to the microprocessor unit. The mixing of the amplified, body-scattered signal and a reference signal (7-10 mW) in the double-balanced mixer produces low-frequency signals resulting from motion due to breathing and heart motion within the body. This output from the mixer is amplified by an operational amplifier and then passed through a low-pass filter (4 Hz cutoff) before reaching a recorder.

Measured Heart and Breathing Signals

Recordings of the heart and breathing signals of several persons were taken under different conditions. However, only a few of them are presented here for illustration. Fig. 4 shows the measured heart and breathing signals of a human subject lying on the ground at a distance of 30 m with a 4.5 mW, 10 GHz microwave beam aimed at him. The top graph in Fig. 4 shows the breathing signal superimposed upon the heart signal when the human subject was lying on the ground in a face-up position with the body perpendicular to the microwave beam. The middle graph in Fig. 4 shows only the heart signal when the subject was holding his breath. The bottom graph in Fig. 4 shows the background noise. The results of Fig. 4 indicate satisfactory performance of the system in detecting the heart and breathing signals of human subjects lying on the ground at a distance of 30 m or farther.

We have also studied the effect of clothing on the system performance by repeating the experiment with different clothing on the subject, e.g., up to four layers of very thick jackets. The effect of the clothing over the sensitivity of the system was found to be insignificant. Furthermore, the polarization effect of the microwave signal on the system performance was also investigated. When circular, linear-vertical, and linear-horizontal polarization were employed, the system sensitivity was found to be of the same order in all three cases.

breathing signals
lying on the ground with face up (body perpendicular to the beam)

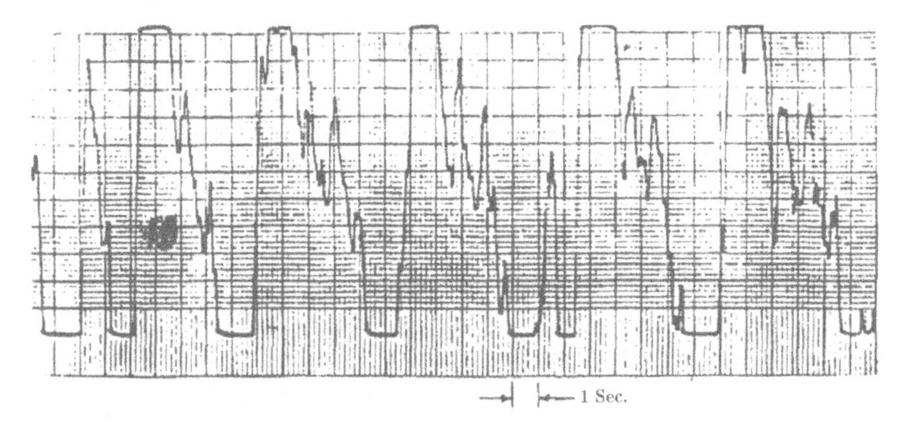

heart signals (holding the breath)
lying on the ground with face up (body perpendicular to the beam)

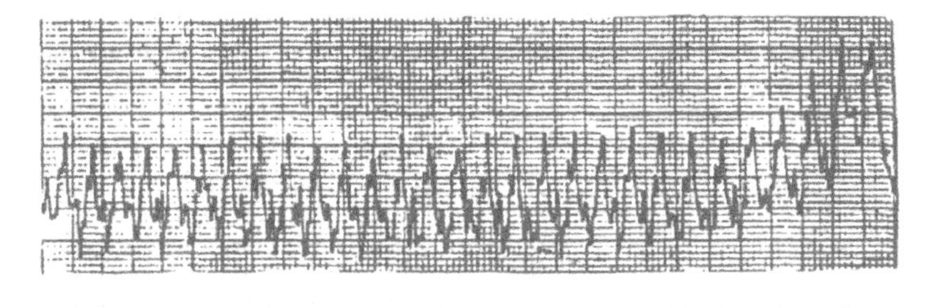

background noise

Fig. 4 Heart and breathing signals of a human subject lying on the ground at a distance of 30 m measured with a 4.5 mW, 10 GHz microwave beam.

Reprinted from "An X-band microwave life-detection system" by Kun-Mu Chen, D. Misra, H. Wang, H. R. Chuang and E. Postow, IEEE Trans. on Biomedical Engineering, Vol. BME-33, No. 7, pp 697-701, July 1986. (©1986 IEEE.)

We have also used this system to detect the heart and breathing signals of a human subject located behind a barrier with success. Fig. 5 shows the measured heart and breathing signals of a human subject sitting behind a dry 15.24 cm (6 in) cinder block wall at a distance of 0.6, 2, or 3 m. The antenna was placed close to the other side of the wall and energized to radiate 20 mW at 10 GHz. It is observed from Fig. 5 that the heart and breathing signals were clearly detected at all three distances. The background noise in each case is also shown in the figure. The results of Fig. 5 indicate that the microwave beam can penetrate the wall and a satisfactory detection of the heart and breathing signals of human subjects behind the wall is possible. We have repeated the experiment by moving both the system and the human subject away from the wall. It was found that the system could perform satisfactorily even when the human subject was 5 m away from the wall while the antenna on the other side was about 3 m from the wall. If the antenna was moved further from the wall, system performance was affected by movement of the system operator.

This life-detection system can be easily modified to produce a device for monitoring the breathing and heartbeat of a patient in a clinic. To conduct such an experiment, a metallic wire-mesh chamber with the dimensions of $2.5 \times 1 \times 0.8$ m was constructed as shown in Fig. 6. The antenna of the system was replaced by an open-ended waveguide which was mounted on a wall of the chamber. A microwave signal of 100 μW at 10 GHz was radiated into the chamber through the waveguide. A human subject was lying inside the chamber in various positions, face up, or lying on his right or left shoulder. The measured heart and breathing signals of the subject lying in these three positions are shown in Fig. 6. It is observed that a clear detection of the heart and breathing signals can be achieved. It is noted that since the microwave field is confined inside a metallic chamber and the environmental noise is minimal, only very low power microwave radiation is needed for this purpose.

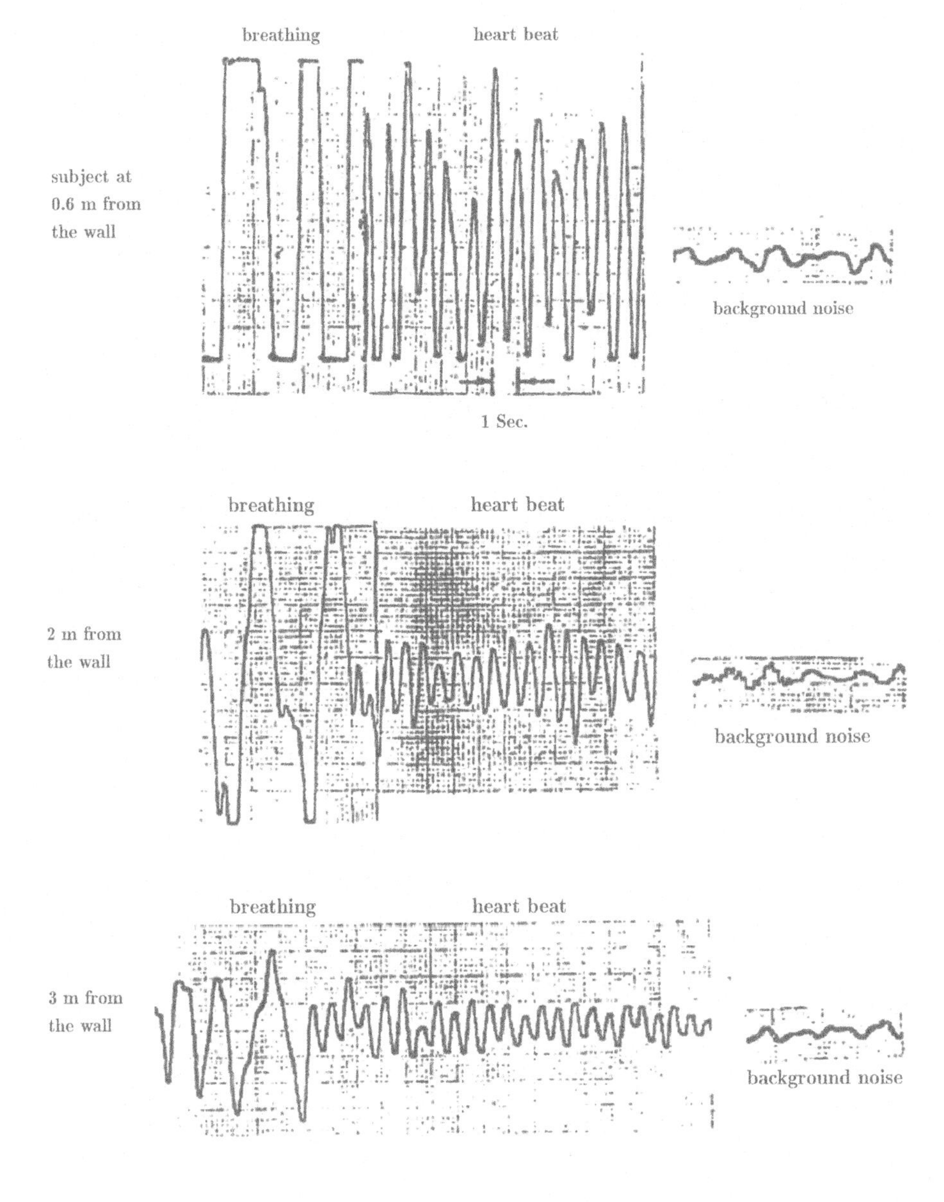

Fig. 5 Measured breathing and heart signals from a human subject sitting behind a cinder block wall (15.24 cm thick) at various distances. The antenna of the life-detection system was located on the other side of the wall and it radiated a power of about 20 mW at 10 GHz.

Reprinted from "An X-band microwave life-detection system" by Kun-Mu Chen, D. Misra, H. Wang, H. R. Chuang and E. Postow, IEEE Trans. on Biomedical Engineering, Vol. B¬¬¬ME-33, No. 7, pp 697-701, July 1986. (©1986 IEEE.)

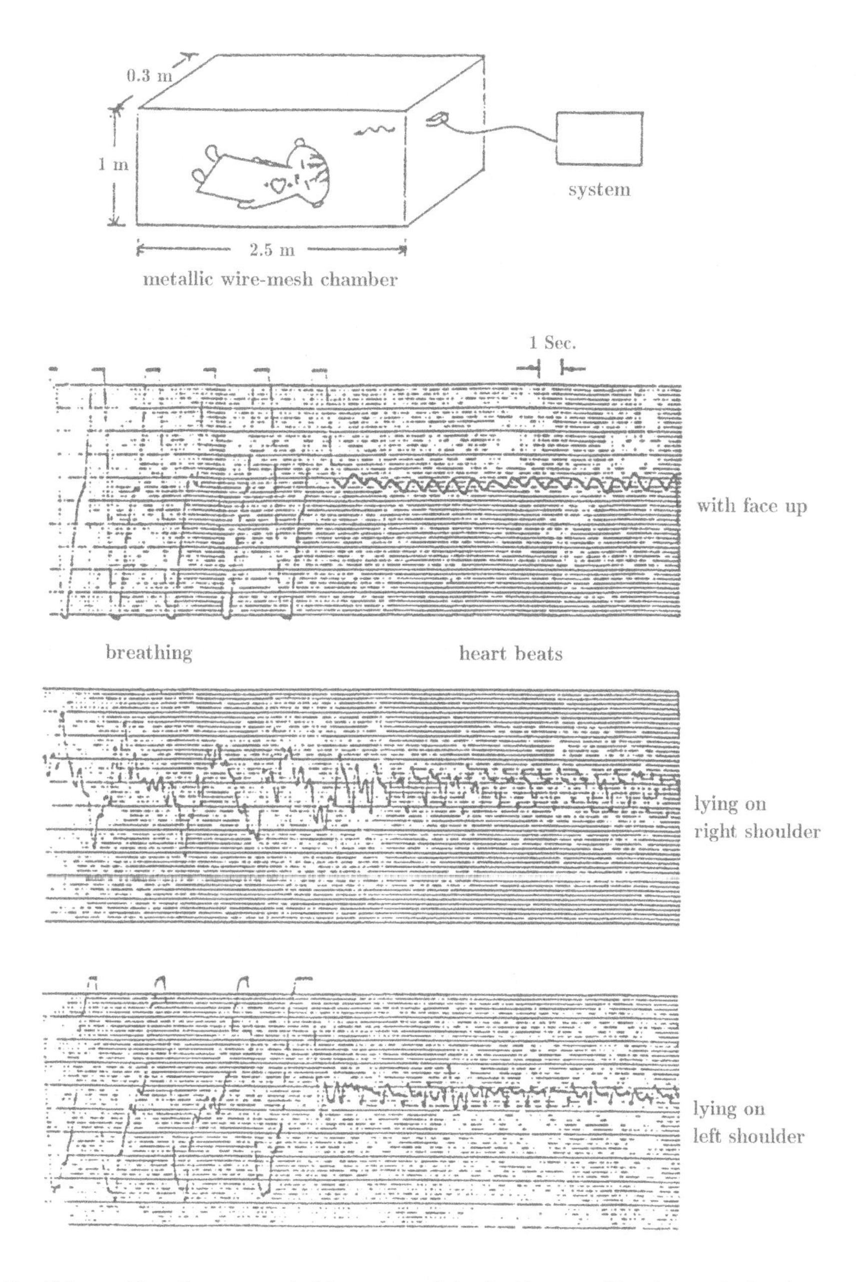

Fig. 6 Breathing and heartbeats recorded for a person lying inside a metallic wire-mesh chamber with dimensions 2.5 × 1.0 × 0.8 m. The body as parallel to the radiation beam, with the head away from the antenna. Transmitted power was about 100 μW.

AN L-BAND MICROWAVE LIFE-DETECTION SYSTEM

As described in the preceding section, the *X*-ban microwave life-detection system can be used to detect breathing and heartbeat of a human subject located behind a brick wall of about 15 cm. However, if the wall became thicker, the detection became difficult with the *X*-band system. An *L*-band (2GHz) system was specially designed for the purpose of detecting breathing and heartbeat of a human subject who was located behind a very thick wall or buried under a thick layer of rubble.

The *L*-band system has the essentially same circuit arrangement as that of the *X*-band system, with the exception that all the components are larger because of the lower operation frequency. Since the *L*-band system operates at a much lower frequency than that of the *X*-band system, its microwave beam is more penetrating.

To test the performance of the *L*-band life-detection system, two experimental setups depicted in Fig. 7 have been used. The first setup shown in Fig. 7a consisted of a brick wall (1 m wide and 1.4 m high) of various thicknesses lined with microwave absorbers along the edge. A human subject sat behind the brick wall within a distance of 0.3 to 0.6 m. The antenna of the life-detection system was placed close to the other side of the brick wall. The second setup shown in Fig.7b simulated a situation where a human subject was trapped under a thick layer of rubble. In this setup, various layers of bricks were laid on a wooden frame which formed a cavity for a human subject to lie down in it. Microwave absorbers were used to line the sides of this structure to prevent the microwave scattering through the sides of the brick structure. The antenna of the life-detection system was placed on the top of the brick structure aiming at the human subject under the bricks.

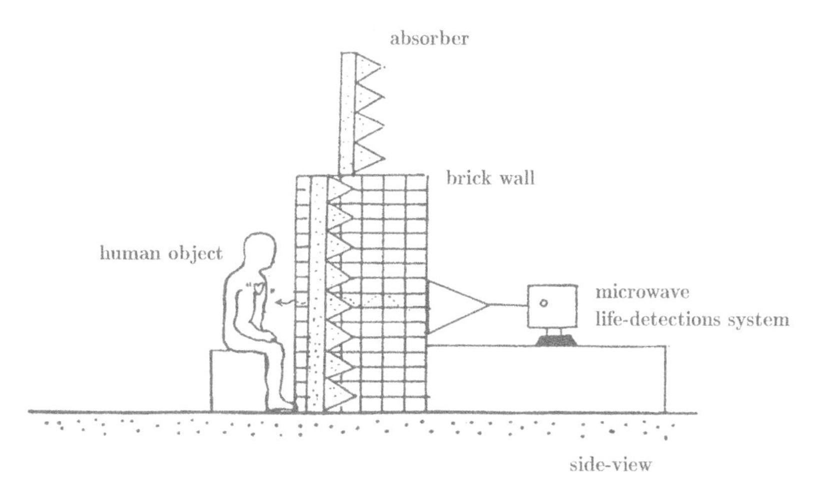

Fig. 7a

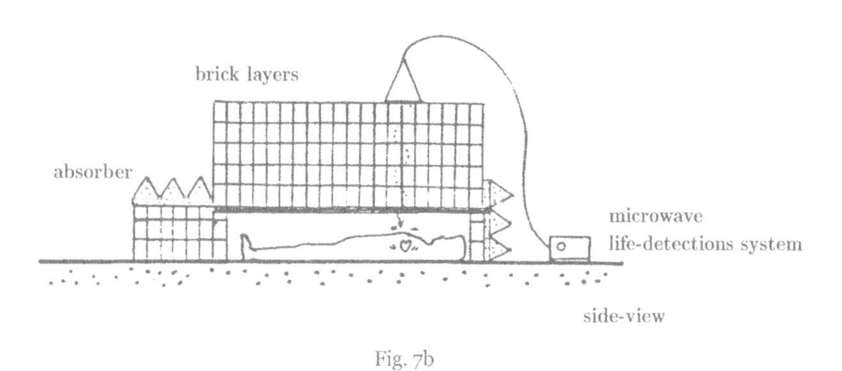

Fig. 7b

Fig. 7 Experimental setups for the measurement of heart and breathing signals of a human subject located behind or under a thick layer of bricks using the L-band (2 GHz) Microwave life-detection system.

Typical measured results on the heart and breathing signals of a human subject behind or under a thick layer of barrier are shown in Fig. 8. This figure shows the heart and breathing signals of a human subject lying with face-up or face-down position under six layers (52 cm) of dry bricks measured by the 2 GHz life-detection system. In these recorded graphs, the breathing signal, the heart signal (the subject holding his breath) and the background noise were included. It is observed that both the heart and breathing signals were clearly detected. These results demonstrate the feasibility of monitoring the physiological signs of human subjects through a thick barrier with an EM radiation with a frequency in the *L*-band or lower range.

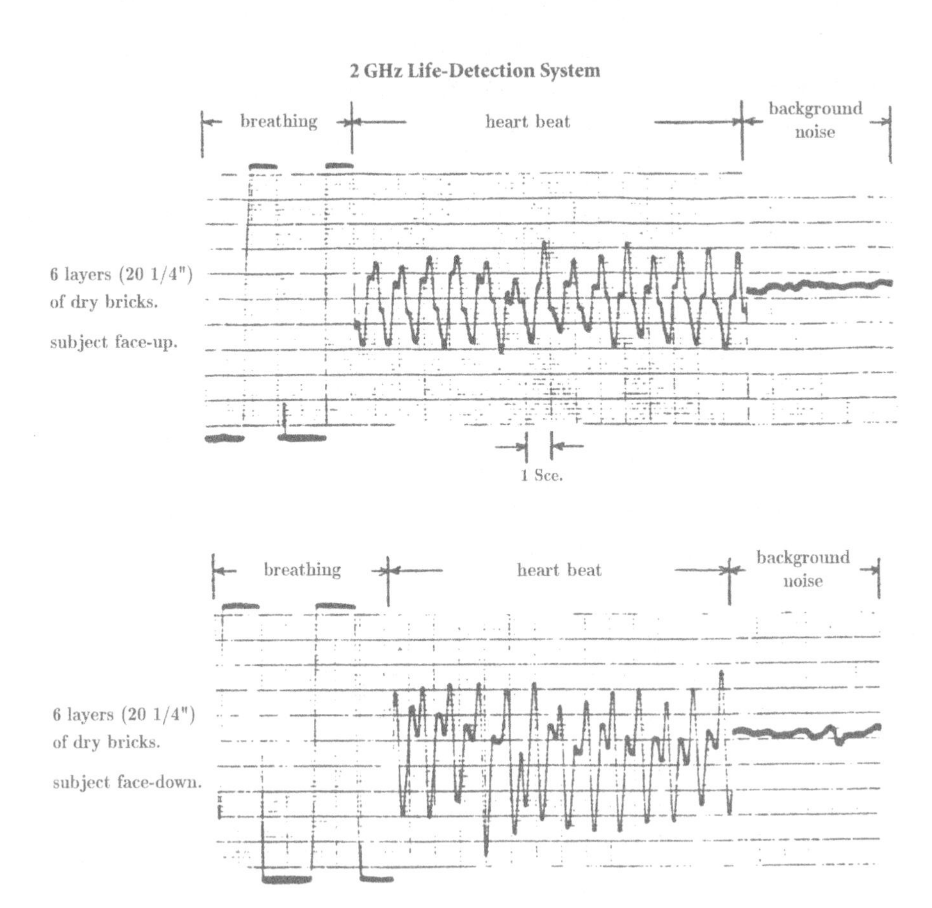

Fig. 8 Heart and breathing signals of a human subjecy, lying with face-up or face-down position under 6 layers of bricks, measured by the 2 GHz life-detection system.

 R.F. Life-Detection Systems for Searching Human Subjects Under Earthquake Rubble or Behind Barrier

In the preceding section, an X-band (10 GHz) and a L-band (2 GHz) microwave life-detection system for the remote sensing of the breathing and heartbeats of a human subject at a distance or behind a barriers were described. Those systems were found to be ineffective if a human subject is covered by a very thick layer of debris such as the earthquake rubble because a microwave radiation at X-band or L-band can not penetrate deep into debris.

Existing methods for searching and rescuing human victims buried under earthquake rubble or collapsed building debris are the utilization of dogs, or seismic or optical devices. These existing devices are not effective if the rubble or debris covering the human victims is thicker than a few feet, especially for the case when the victims are completely trapped or too weak to respond to the signal sent by the rescuers. Thus, there is great demand for constructing a new sensitive life-detection system which can be used to locate human victims trapped deep under earthquake rubble or collapsed building debris. Especially, the system needs to be sensitive enough to detect the breathing and heartbeat signals of passive victims who are completely trapped or too weak to respond to the existing seismic detection system.

A sensitive life-detection system for such purpose was constructed recently by us at Michigan Sate University [4]. This system operating at 450 MHz or at 1150 MHz will be described in this section.

The basic physical principle for the operation of a microwave life-detection system is rather simple. When an EM wave beam of appropriate frequency (L or S band) is aimed at a pile of earthquake rubble or collapsed building debris under which a human subject is buried, the EM wave beam can penetrate through the rubble or the debris to reach the subject. When the human subject is illuminated by the EM wave beam, the reflected wave from the subject will be modulated by the subject's body movements, which include the breathing and the heartbeat. If the reflected wave from the stationary background can be cancelled and the reflected wave from the subject's body is properly demodulated, the breathing and heartbeat signals of the subject can be extracted. Thus, a human subject buried under the rubble or the debris can be located.

The system operating at 450 MHz was constructed first. This system was tested on simulated earthquake rubble constructed at the Electromagnetics Laboratory at Michigan State University, and it was also tested in a field test using realistic earthquake rubble consisted of layers of reinforced concrete slabs with imbedded metallic wire mesh at a test site in Rockville, MD, with the cooperation of the Maryland Task Force of the Federal Emergency Management Agency (FEMA). The results of these tests will be described. The second system operating at 1150 MHz was constructed after the field test at Rockville, MD. In that field test, it was found that an EM wave of 450 MHz is difficult to penetrate layers of reinforced concrete slabs with imbedded metallic wire of 4-in spacing. Through a series of experiment, we selected the operating frequency of 1150 MHz for the second system with the goal of penetrating such earthquake rubble. After the construction of the 450-MHz and the 1150-MHz systems and an extensive series of experiments, we found that an EM wave of 1150 MHz can penetrate a rubble with layers of reinforced concrete slabs with metallic wire mesh easier than that of 450 MHz. However, an EM wave of 450 MHz may penetrate deeper into a rubble without metallic wire mesh than that of 1150 MHz.

The R.F. life-detection system we constructed has four major components: 1) a microwave circuit system which generates, amplifies, and distributes microwave signals to various microwave components; 2) a microprocessor-controlled clutter-cancellation system which creates an optimal signal to cancel the clutter from the rubble and the background; 3) a dual-antenna system which consists of two separate antennas energized sequentially; and 4) a laptop computer which controls the microprocessors and acts as the monitor for the output signal. The system is operated by a portable battery unit.

Both the 450-MHz and the 1150-MHz systems are working well for various types of earthquake rubble and collapsed building debris. They can detect the breathing and heartbeat signals of trapped human subjects buried under a rubble of up to 10-ft thickness.

CIRCUIT DESCRIPTION OF THE SYSTEM

The basic circuit structures of the 450-MHz and the 1150-MHz microwave life-detection systems are quite similar and they are operated based on the same physical principle. In this section, only the circuit structure of the 1150-MHz system will be described, while that of the 450-MHz system is very similar.

The schematic diagram of the 1150-MHz microwave life-detection system is shown in Fig. 9. A phase-locked oscillator generates a very stable EM wave at 1150 MHz with an output power of 400mW (25.6 dBm). This wave is fed through a 10-dB directional coupler and a circulator before reaching a radio-frequency (RF) switch, which energized the dual antenna system sequentially. The 10-dB directional coupler branches out one-tenth of the wave (40 mW) which is then divided equally by a 3-dB directional coupler. One output of the 3-dB directional coupler (20 mW) drives the clutter cancellation circuit and the other output (20 mW) serves as a local reference signal for the double-balanced mixer.

The wave radiated by an antenna penetrates the earthquake rubble to reach a buried human subject. The reflected wave received by the same antenna consists of a large reflected wave (clutter) from the rubble and a small reflected wave from the subject's body. The large clutter from the rubble can be cancelled by a clutter canceling signal. However, the small reflected wave from the subject's body cannot be cancelled by a pure sinusoidal, canceling signal because it is modulated by the subject's motions. The dual-antenna system has two antennas, which are energized sequentially by an electronic switch. Each antenna acts independently and the final outputs from these two antennas are combined in some signal processing schemes to reduce the background noise. This part will be elaborated later.

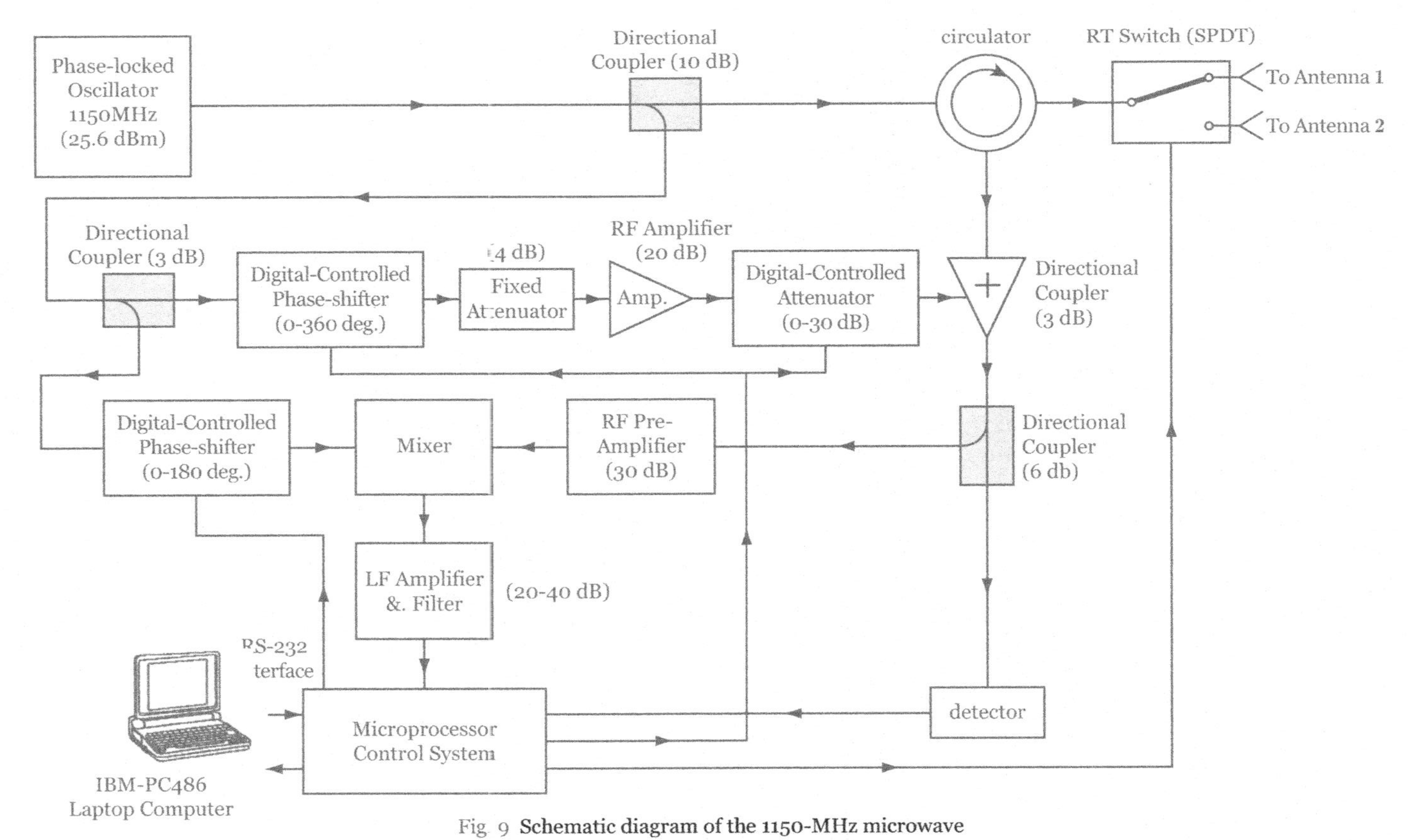

Fig. 9 **Schematic diagram of the 1150-MHz microwave**

Reprinted from "Microwave life detection system for searching human subjects under earthquake rubble or behind barrier" by Kun-Mu Chen, Y. Huang, J. Zhang and A. Norman, IEEE Trans. on Biomedical Engineering, Vol. 27, No. 1, pp. 105-114, Jan. 2000. (©2000 IEEE.)

The clutter cancellation circuit consists of a digitally controlled phase-shifter (0°–360°), a fixed attenuator (4 dB), a RF amplifier (20 dB), and a digitally controlled attenuator (0–30 dB). The output of the clutter cancellation circuit is automatically adjusted to be of equal amplitude and opposite phase as that of the clutter from the rubble. Thus, when the output of the clutter cancellation circuit is combined with the received signal from the antenna, via the circulator, in a 3-dB directional coupler, the large clutter from the rubble is completely canceled, and the output of the 3-dB directional coupler consists only of the small reflected wave from the subjects body. This output of the 3-dB directional coupler is passed through a 6-dB directional coupler. The 1/4 of this output is amplified by a RF preamplifier (30 dB) and then mixed with a local reference signal in a double-balanced mixer. The other 3/4 of the output is detected by a microwave detector to provide a dc voltage, which serves as the indicator for the degree of the clutter cancellation. When the settings of the digitally controlled phase-shifter and attenuator are swept by the microprocessor control system, the output of the microwave detector varies accordingly. The minimum detector reading corresponds to the right settings for the digitally controlled phase-shifter and attenuator. These settings will be fixed for subsequent measurements.

At the double-balanced mixer, the amplified signal of the reflected wave from the subject's body is mixed with a local reference signal. The phase of the local reference signal is controlled by another digitally controlled phase-shifter (0°-180°) for an optimal output from the mixer. (This function will be elaborated on later.) The output of the mixer consists of the breathing and heartbeat signals of the human subject plus unavoidable noise. This output is fed through a low-frequency (LF) amplifier (20–40 dB) and a bandpass filter (0.1–4 Hz) before being displayed on the monitor of a laptop computer.

The function of a digitally controlled phase-shifter (0°-180°) installed in front of the local reference signal port of the double-balanced mixer to control the phase of the local reference signal for the purpose of increasing the system sensitivity is explained below.

As mentioned before, the reflected signal from the human subject after amplification by the pre-amplifier is mixed with the local reference signal in the double-balanced-mixer. The local reference signal is assumed to be $A_L \cos(\omega t + \phi_L)$ where A_L and ϕ_L are the amplitude and the phase, respectively. While the other input to the mixer, the reflected signal from the human subject, is assumed to be $A_r \cos(\omega t + \phi_E + \Delta\phi(t))$ where A_r and ϕ_E are the amplitude and the phase, respectively, and $\Delta\phi$(t) is the phase modulation due to the body movement of the human subject. ω is the angular frequency and t is the time. When these two inputs are mixed in the double-balanced mixer, the output of the mixer will be $A_L A_r \cos(\phi_L - \phi_E - \Delta\phi(t))$.

From this expression of the mixer output, it is easy to see that

$$\text{If } \phi_L - \phi_E = (n + 1/2)\pi,\ n = 0,1,2,... \tag{4.8}$$

the system has a maximum sensitivity;

and

$$\text{If } \phi_L - \phi_E = \pm n\pi,\ n = 0,1,2,... \tag{4.9}$$

the system has a minimum sensitivity,

because $(\partial / \partial\Delta\phi(t))\cos(\phi_L - \phi_E - \Delta\phi(t)) = -\sin(\phi_L - \phi_E - \Delta\phi(t))$. $\Delta\phi$(t) is usually a small phase angle perturbation created by the body movement of the human subject. ϕ_E is the constant phase associated with the reflected signal from the human subject and it cannot be changed. ϕ_L is the phase of the local reference signal and it can be controlled by the digitally controlled phase-shifter (0°-180°). In the operation, the phase-shifter will automatically shift in such a way that $\phi_L - \phi_E$ is nearly $(n+1/2)\pi$ to attain a maximum system sensitivity.

ANTENNA SYSTEM

We have designed and constructed three types of antennas for the microwave life-detection system. They are: 1) the reflector antenna; 2) the patch antenna; and 3) the probe antenna. Each antenna simultaneously acts as the radiating element and the receiving element. It radiates EM wave through the earthquake rubble to reach the trapped human subjects and at the same time it receives the reflected EM wave from the rubble and the human subjects. The antenna can perform two functions simultaneously with the help of a circulator, which separates the radiating EM wave from the received EM wave.

The reflector antenna was constructed with two aluminum plates as the reflectors and an adjustable dipole antenna as the driving element. The two aluminum plates with the dimensions of 21 in × 11 in form a corner reflector with the dipole antenna as its primary radiator. The angle between the two aluminum plates is adjustable and they are folded together when it is not used. The dipole antenna is a conventional, half-wavelength electric dipole. The reflector antenna is a simple, lightweight, and ragged structure and it performs very well in the most of situations.

The gain of the reflector antenna is difficult to define and measure because the antenna is placed directly over a rubble pile and the scattered field of the antenna is strongly dependent on the nature of the rubble material.

A patch antenna was constructed for radiating and receiving EM wave for the microwave life-detection system. The patch antenna consists of an aluminum ground plane, which is supported by four legs and a strip plate of about a half-wavelength, which is attached to the ground plane and fed by a coaxial line. The strip plate is insulated from the ground plane. The coaxial cable is attached to the ground plane through a connector.

The performance of the patch antenna is not better than that of the reflector antenna. It only serves as alternative type of antenna and may be useful in some situations.

A probe antenna was designed to insert through boreholes or naturally occurring fissures into the earthquake rubble to seek for the trapped victims. Physically, a probe antenna should have a cylindrical wire structure and its radius be kept as small as possible. We have designed a probe antenna, which is essentially a sleeve antenna, as shown in Fig. 10.

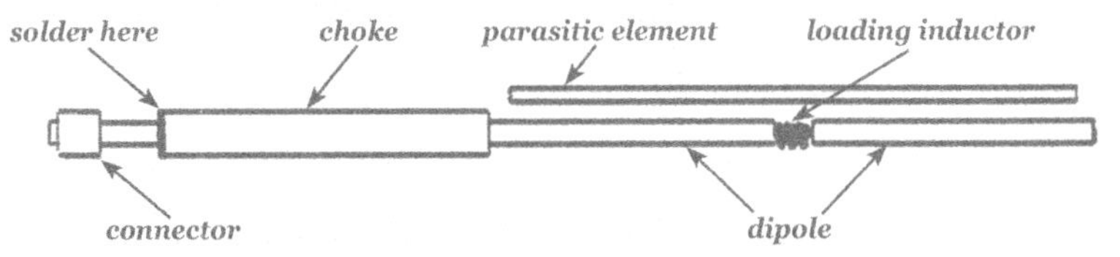

Fig. 10 Probe antenna for the life-detection system.

The radiating element is a half-wavelength dipole, which is loaded with an inductor at the center. The inductance of the inductor was determined numerically in the design. One half of the dipole is connected to the center conductor of the coaxial cable via the inductor. The other half of the dipole is a quarter-wavelength section of the outer surface of the coaxial cable. A quarter-wavelength choke, which is cylindrical tubing of larger radius than that of the coaxial cable, is soldered to the coaxial cable at one end and kept open at the other end. This choke is acting as a shorted, quarter-wavelength transmission line, which provides very high input impedance at the end point of the radiating dipole. Thus, this choke will stop the unbalanced current leaking to the outer surface of the connecting cable. A parasitic element, a wire of slightly shorter than half-wavelength, is placed next to the radiating dipole to increase the bandwidth of the antenna. The selection of dimensions of the parasitic element was made empirically through an experiment with a network analyzer. The whole structure of the probe antenna is encased in a rugged plastic tubing.

The dual antenna system has two antennas, which are energized sequentially by an electronically controlled microwave single-pole double-throw (SPDT) switch. The SPDT switch turns on and off at a frequency of 100 Hz which is much higher than the frequency range of the breathing and heartbeat signals between 0.2 Hz and 3 Hz. Thus, we can consider that the two antennas essentially sample their respective objects at the same time. In this dual-antenna system, the two antenna channels are completely independent.

EXPERIMENTAL RESULTS

The 450- and 1150-MHz microwave life-detection systems were tested in a simulated earthquake rubble constructed at the Electromagnetics Laboratory of Michigan State University. The 450-MHz system was also tested in a field-test with realistic rubble conducted at Montgomery County, Rockville, MD, with the cooperation of Maryland Task Force of FEMA. Typical experimental results of these systems are summarized here.

A. Experimental Results Obtained with the 450-MHz System at a Simulated Rubble in MSU Laboratory

The 450-MHz microwave life-detection system was tested in simulated rubble constructed in the Electromagnetics Laboratory of Michigan State University. The rubble is depicted in Fig. 11. It was constructed with bricks, cinder blocks, and steel re-bars. The dimensions of the rubble was about 5 ft wide, 6 ft long, and 6 ft high. Two layers of steel re-bars separated by 8 in are placed perpendicularly through bricks as shown in Fig. 11. A human subject to be tested can lie down in the cavity at the bottom of the rubble. A reflector antenna or a patch antenna can be placed on the top of the rubble, while a probe antenna can penetrate into the rubble through a hole in the rubble.

Typical experimental results of the breathing and heartbeat signals of a human subject lying in the rubble cavity obtained with the 450-MHz system are shown in Figs. 12 and 13.

Fig. 12 shows a breathing signal superimposed with a heartbeat signal recorded for a female human subject. A reflector antenna was used and the radiated power was about 300 mW. The upper graph is the time domain measured signal and the lower graph is the fast Fourier transform (FFT) of the time-domain signal, which shows the frequency components of the time-domain signal. The upper graph clearly shows the breathing and heartbeat signals. The frequency domain FFT results show that the time-domain signal has a breathing signal of 0.3 Hz (the dominant peak) and a heartbeat signal of 1.36 Hz (the second largest peak). The other peak at 0.6 Hz (the third largest peak) is the second harmonic of the breathing signal. Other small peaks are due to noise or harmonics of the breathing and heartbeat signals. From a signal as shown in Fig. 12, it is easy to identify the breathing and heartbeat signals from either the time-domain signal or the frequency domain FFT results, and a buried human subject is easily detected.

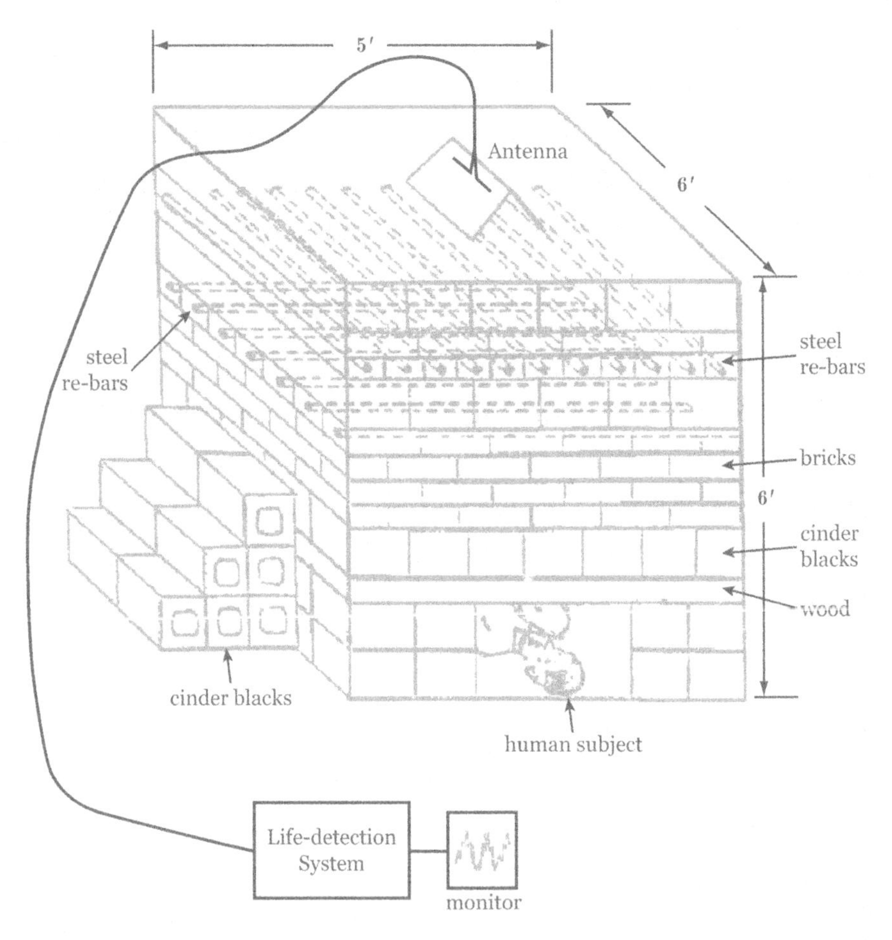

Fig. 11 **Simulated earthquake rubble constructed at the Electromagnetics Laboratory of Michigan State University.**

Reprinted from "Microwave life detection system for searching human subjects under earthquake rubble or behind barrier" by Kun-Mu Chen, Y. Huang, J. Zhang and A. Norman, IEEE Trans. on Biomedical Engineering, Vol. 27, No. 1, pp. 105-114, Jan. 2000. (©2000 IEEE.)

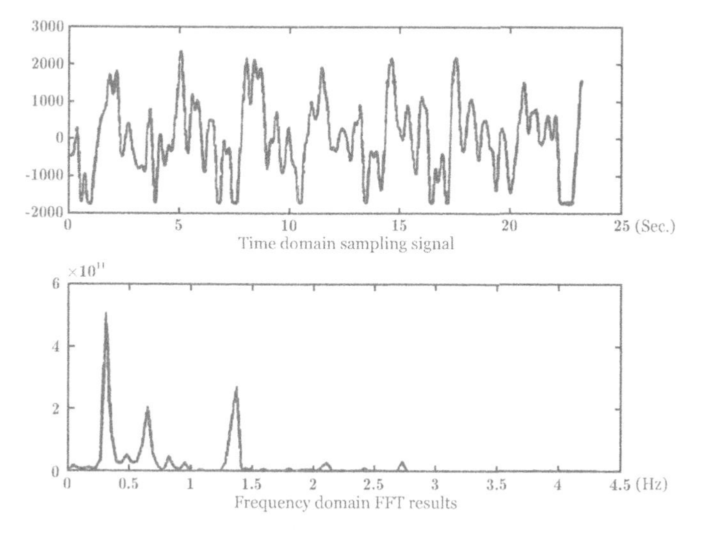

Fig. 12 Breathing and heartbeat signals of a female human subject recorded at MSU simulated rubble. A reflector antenna was placed on the top of the rubble and the female human subject was lying inside the rubble cavity. The radiated power is about 300 mW. The 450-MHz life-detection system was used.

Reprinted from "Microwave life detection system for searching human subjects under earthquake rubble or behind barrier" by Kun-Mu Chen, Y. Huang, J. Zhang and A. Norman, IEEE Trans. on Biomedical Engineering, Vol. 27, No. 1, pp. 105-114, Jan. 2000. (©2000 IEEE.)

Fig. 13 shows the same measurement conducted on the same subject when she was holding her breath. The time-domain signal (upper graph) shows only the heartbeat signal and the frequency domain FFT results (lower graph) shows only a single dominant peak of heartbeat signal at 1.36 Hz. Other small peaks are probably due to noise. It is noted that when the signals of Figs. 12 and 13 are compared, the amplitude of heartbeat signal is found to be significantly smaller than that of the breathing signal as expected.

Fig. 14 shows the background noise recorded when no human subject was in the rubble cavity. It is noted that the amplitude of the noise is lower than that of the breathing signal and the noise has wide spread frequency components as indicated in its FFT results. It is easy to distinguish the noise from the breathing and heartbeat signals from the amplitude and the frequency contents of the recorded signals.

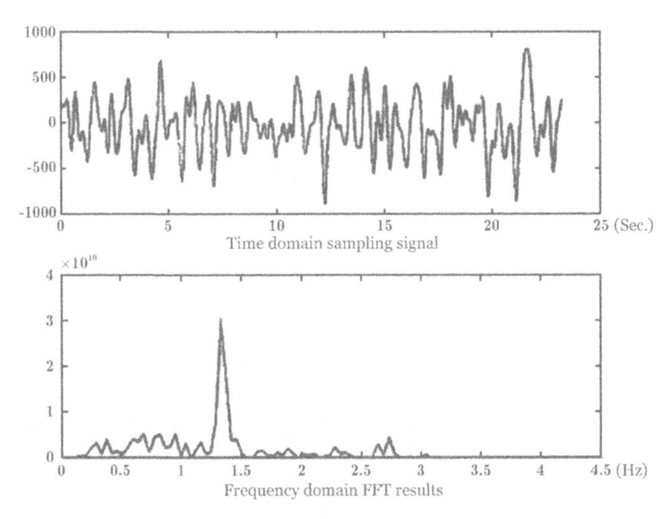

Fig. 13 **Heartbeat signal of a female human subject recorded at MSU simulated rubble. A reflector antenna was placed at the top of the rubble and the female human subject lying inside the rubble cavity was holding her breath. The 450-MHz life-detection system was used.**

Reprinted from "Microwave life detection system for searching human subjects under earthquake rubble or behind barrier" by Kun-Mu Chen, Y. Huang, J. Zhang and A. Norman, IEEE Trans. on Biomedical Engineering, Vol. 27, No. 1, pp. 105-114, Jan. 2000. (©2000 IEEE.)

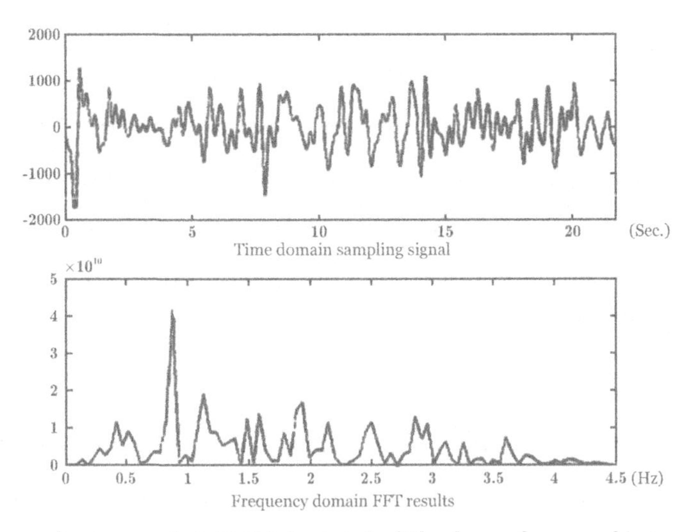

Fig. 14 **Background noise recorded at MSU simulated rubble when no human subject was inside the rubble cavity. The 450-MHz life-detection system was used.**

Reprinted from "Microwave life detection system for searching human subjects under earthquake rubble or behind barrier" by Kun-Mu Chen, Y. Huang, J. Zhang and A. Norman, IEEE Trans. on Biomedical Engineering, Vol. 27, No. 1, pp. 105-114, Jan. 2000. (©2000 IEEE.)

B. *Experimental Results Obtained with the 450-MHz System in the Field Test Conducted at Rockville, MD*

On July 5–7, 1995, a field test managed by Maryland FEMA Task Force was conducted at its Montgomery County training ground. They constructed three rubble structures using reinforced concrete slabs and double-T structures for the test. The first rubble having a height of 6 ft was constructed with seven layers of reinforced concrete slabs placed on the top of a double-T structure, simulating a collapsed seven-story building. The second rubble having a height of 9 ft was constructed with six layers of reinforced concrete slabs on the top of two double-T structures. The third rubble was constructed with pieces of reinforced concrete blocks piled on the top of a reinforced concrete pipe, which had a diameter of about 2 ft, and a fine metallic wire mesh imbedded. Also there was wet soil mixed in the rubble and the rubble was directly under large trees. The height of this rubble was about 9 ft. This rubble used in the field test is much more difficult for an EM wave of 450 MHz to penetrate than the simulated rubble used in MSU Laboratory because of the dimensions and the contents of the rubble and many layers of metallic wire mesh present in the rubble.

Many experimental results were recorded for various conditions by changing the locations of antenna and human subjects and using different rubble. However, only some of the test results measured at the second rubble, as depicted in Fig. 15, will be presented here for brevity.

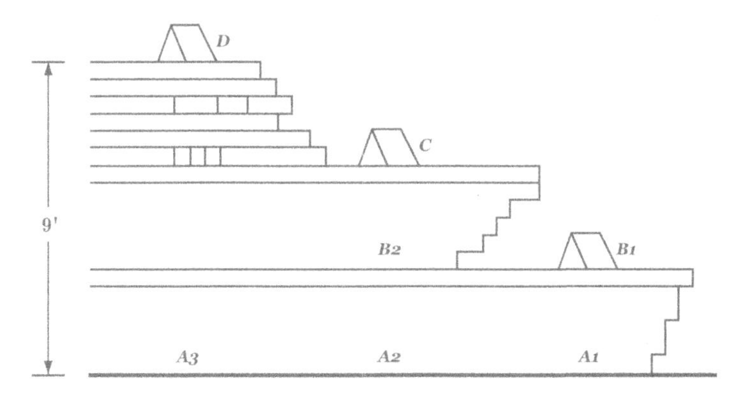

Fig. 15 The second earthquake rubble constructed at Montgomery County, Rockville, MD.

Fig. 16 shows the results of a test conducted at the second rubble, with the reflector antenna placed at location *B1* and a human subject lying at location *A1*. In this case, the EM wave needed to penetrate only one double-T structure, therefore a very strong breathing signal was recorded. The time-domain signal showing a strong breathing signal was recorded. The time-domain signal shows a strong breathing signal (over the scale) and its FFT results show a single dominant peak at 0.26 Hz. Because of the overwhelmed breathing signal the heartbeat signal was overshadowed.

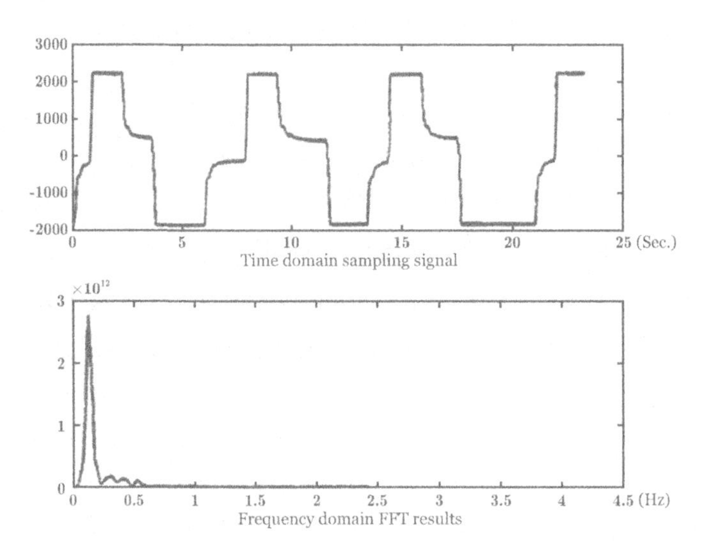

Fig. 16 **Breathing signal of a human recorded at the second rubble, with a reflector antenna at location *B1* and the human subject lying at location *A1*. The 450-MHz life-detection system was used.**

Reprinted from "Microwave life detection system for searching human subjects under earthquake rubble or behind barrier" by Kun-Mu Chen, Y. Huang, J. Zhang and A. Norman, IEEE Trans. on Biomedical Engineering, Vol. 27, No. 1, pp. 105-114, Jan. 2000. (©2000 IEEE.)

Fig. 17 presents the results of a test conducted at the second rubble, with the reflector antenna placed at location *D* and a human subject lying near location *A3*. In this case, the EM wave needed to penetrate six layers of reinforced concrete slabs and two double-T structures. Because of a great depth (about 9 ft) of dense rubble existed between the antenna and the human subject, the magnitude of the received signal was considerably reduced. The time-domain signal shows a distinctive breathing signal and a mixture of heartbeat signal and noise. However, its FFT results clearly identify the breathing signal and a possible heartbeat signal. It is noted that two peaks appeared near 0.2–0.3 Hz may be due to the uneven breathing pattern of the human subject under test.

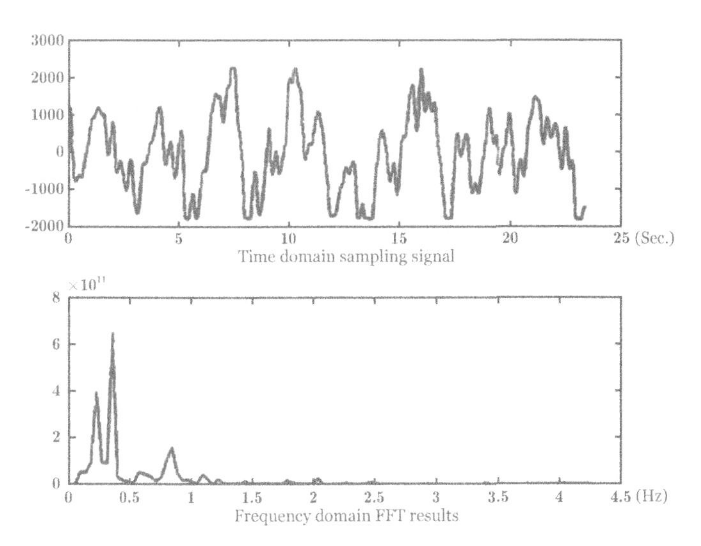

Fig. 17 Breathing and heartbeat signals of a human subject recorded at the second rubble, with a reflector antenna placed at location *D* and the human subject lying near location *A3*. The 450-MHz life-detection system was used.

Reprinted from "Microwave life detection system for searching human subjects under earthquake rubble or behind barrier" by Kun-Mu Chen, Y. Huang, J. Zhang and A. Norman, IEEE Trans. on Biomedical Engineering, Vol. 27, No. 1, pp. 105-114, Jan. 2000. (©2000 IEEE.)

Fig. 18 shows the measured background noise when no human subject was present under the rubble pile. It is noted that the amplitude of the time-domain signal was reduced one order of magnitude from the case when a human subject was present under the rubble pile. The FFT results showed the presence of wide spread frequency components implying a random noise. However, it is pointed out that the two large peaks near 0.3 and 1.3 Hz were also recorded. It is suspected to be contributed by the operator taking the measurement. This may cause misjudgment in the rescue effort. To avoid this problem, when the measured signal is very low and peaks indicating potential breathing or heartbeat signals are present, the operator needs to move around to check his potential interference. This problem can also be mitigated by a dual-antenna system as discussed later.

The performance of the 450-MHz system at the field test was satisfactory. However, it was also found that an EM wave of 450 MHz does not penetrate well a rubble consisting of layers of reinforced concrete slabs with imbedded metallic wire mesh of 4-in spacing. To overcome this

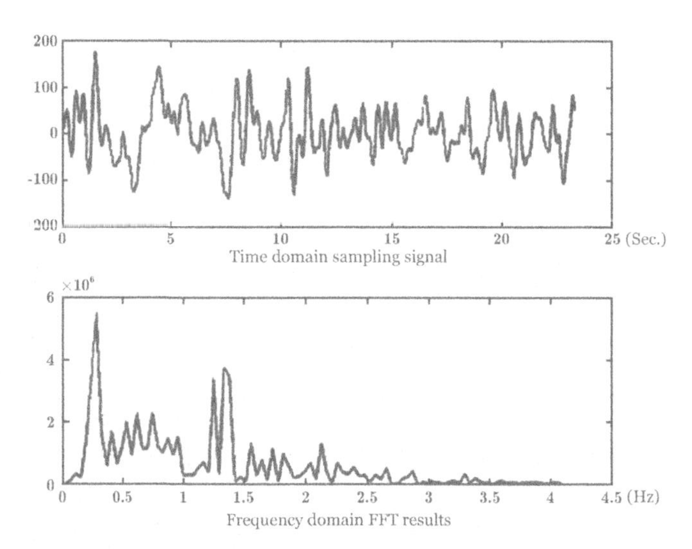

Fig. 18 Background noise recorded when no human subject was inside the rubble. The 450-MHz life-detection system was used.

Reprinted from "Microwave life detection system for searching human subjects under earthquake rubble or behind barrier" by Kun-Mu Chen, Y. Huang, J. Zhang and A. Norman, IEEE Trans. on Biomedical Engineering, Vol. 27, No. 1, pp. 105-114, Jan. 2000. (©2000 IEEE.)

problem, the frequency of the EM wave needs to be raised higher, but not too high to greatly reduce its penetration depth. Through some experiments, we selected the operating frequency of 1150 MHz for our second system. The experimental results of the 1150-MHz system are described in the following section.

C. Experimental Results Obtained with the 1150-MHz System at a Simulated Rubble in MSU Laboratory with Simulated Interference

The performance of the 1150-MHz system is quite similar to that of the 450-MHz system. The main difference is that the EM wave of 1150 MHz can penetrate through a rubble, which consisted of layers of reinforced concrete slabs with imbedded metallic wire mesh of 4-in spacing, better than the EM wave of 450 MHz. That is the 1150-MHz system should be more effective in detecting the breathing and heartbeat signals of trapped victims under the collapse of reinforced concrete and masonry structures than the 450-MHz system. On the other hand, the 450-MHz system may be more effective in searching for trapped victims under a rubble which does not have fine metallic wire mesh imbedded because an EM wave of 450 MHz can penetrate deeper than an EM wave of 1150 MHz in such a medium.

The experimental results obtained with the 1150-MHz system using the MSU simulated rubble on the detection of the breathing and heartbeat signals of human subjects are very similar to that obtained with the 450-MHz system as described earlier and they will not be repeated here. However, we will concentrate on the experimental results using the dual-antenna system of the 1150-MHz system and also on its performance on mitigating the background interference.

In this series of experiments, an artificial breather and an artificial heart which simulate the breathing and the heartbeat signals of a human subject were used as the target. The MSU Laboratory rubble was modified by inserting more steel re-bars to form two layers of metallic wire meshes of 4-in spacing.

The experimental results in the next three figures show only the heartbeat signals created by the artificial heart because the artificial breather was shut off. The thrust of these three figures is to show that the detection of the heartbeat signal can be enhanced if the two sets of signals received by antenna *A* and antenna *B* are crosscorrelated.

Fig. 19 shows the heartbeat signals created by the artificial heart and measured by reflector antenna *A* and reflector antenna *B* which were placed 7 ft directly above the target. The time-domain results of both antennas show the heartbeat signals contaminated by a large noise. Their FFT results also show the presence of a strong noise with spread frequencies. However, when these two sets of signals were crosscorrelated, a distinctive peak of the heartbeat signal at 0.8

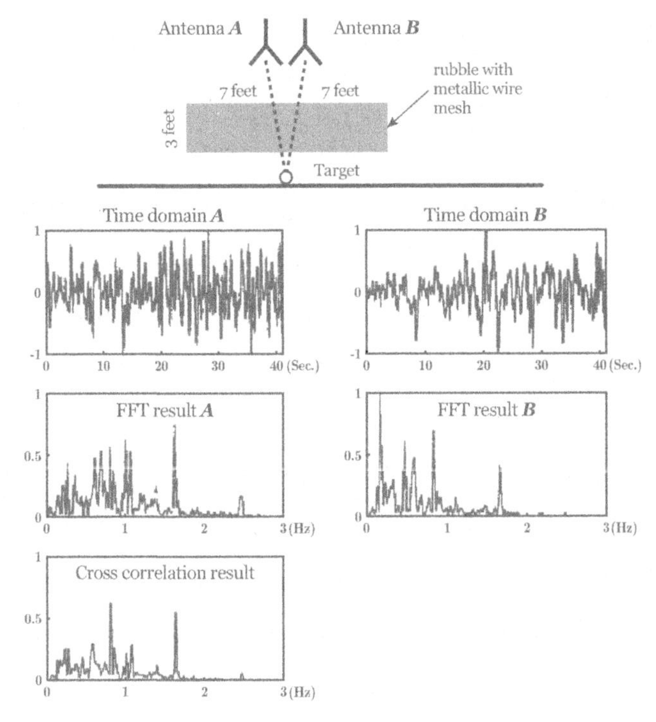

Fig. 19 Heartbeat signals measured by two-reflector antennas arranged symmetrically. Both time-domain and FFT results are shown. The cross-correlation result of the two sets of results shows two peaks representing the heartbeat frequency and its second harmonic. The 1150 -MHz life-detection system was used.

Reprinted from "Microwave life detection system for searching human subjects under earthquake rubble or behind barrier" by Kun-Mu Chen, Y. Huang, J. Zhang and A. Norman, IEEE Trans. on Bio-medical Engineering, Vol. 27, No. 1, pp. 105-114, Jan. 2000. (©2000 IEEE.)

Hz appeared. A second distinctive peak at 1.6 Hz is the second harmonic of the heartbeat signal. It is also observed that the noise measured by both antennas was drastically reduced. From this cross-correlated result, the heartbeat signal was clearly detected.

Fig. 20 shows the heartbeat signals measured by two different types of antennas. The reflector antenna *A* was placed 7 ft above the target and the probe antenna *B* was inserted through the rubble to reach a point 3.5 ft from the target. The time-domain signals measured by both antennas are shown. For this case the FFT results of these two sets of signals both show a distinctive heartbeat signal and its harmonics. When these two sets of signals are crosscorrelated, a more distinctive heartbeat signal at 0.8 Hz and its second harmonic at 1.6 Hz are produced.

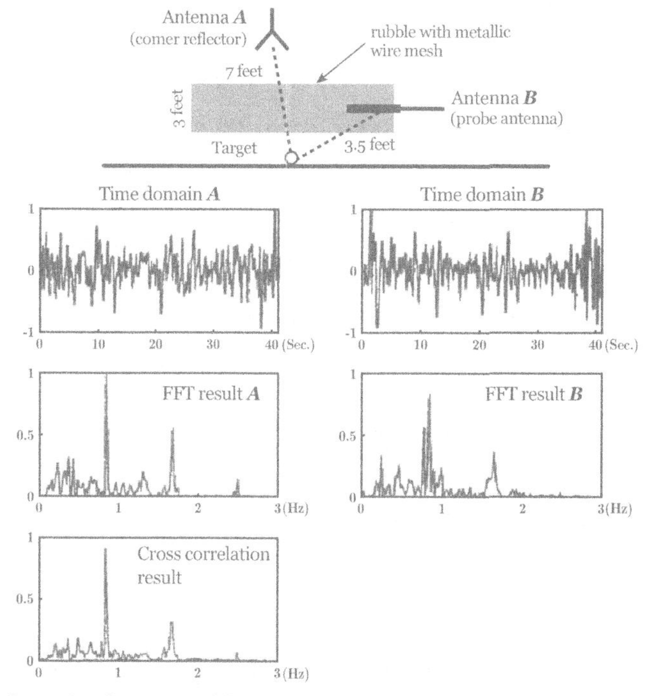

Fig. 20 Heartbeat signals measured by a reflector antenna and a probe antenna. Both time-domain and FFT results are shown. The cross-correlation result of the two sets of results shows the heartbeat frequency and its second harmonic. The 1150-MHz life-detection system was used.

Reprinted from "Microwave life detection system for searching human subjects under earthquake rubble or behind barrier" by Kun-Mu Chen, Y. Huang, J. Zhang and A. Norman, IEEE Trans. on Biomedical Engineering, Vol. 27, No. 1, pp. 105-114, Jan. 2000. (©2000 IEEE.)

Fig. 21 shows the heartbeat signals measured by reflector antenna A and reflector antenna B both placed 7 ft above the target when a human operator was walking near the rubble, about 20 ft from the antenna. The walking human subject created a large interference signal in the outputs of antenna *A* and antenna *B* showing both in their time-domain results and the FFT results. When those two sets of signals were crosscorrelated, the heartbeat signal of 0.8 Hz and its second harmonic of 1.6 Hz appeared while the interference signal nearly disappeared. From this result, we can conclude that the dual-antenna system of the 1150-MHz can be used to reduce the interference noise created by the system operators moving near the rubble as well as the background noise.

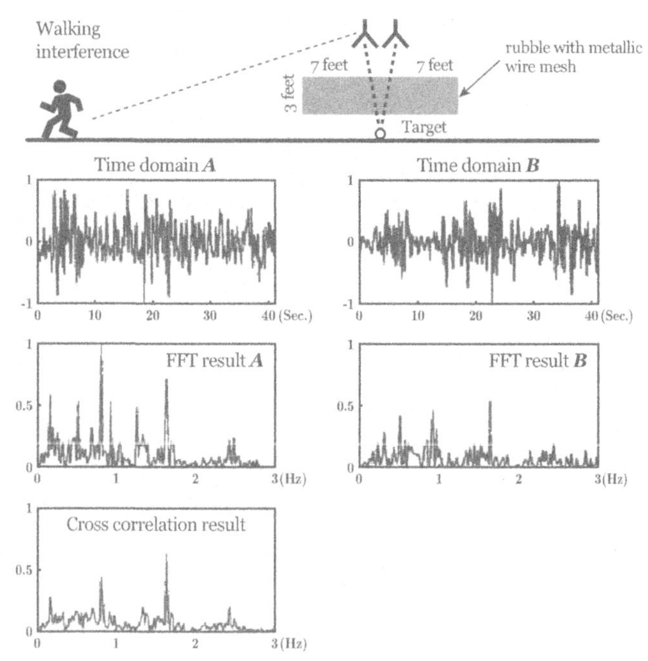

Fig. 21 Heartbeat signals measured by two reflector antennas while a human operator was walking near the rubble. Both time-domain and FFT results are shown. The cross-correlation result of the two sets of results shows the heartbeat frequency and its harmonic, while the interference signal created by the operator nearly disappear. The 1150-MHz life-detection system was used.

Reprinted from "Microwave life detection system for searching human subjects under earthquake rubble or behind barrier" by Kun-Mu Chen, Y. Huang, J. Zhang and A. Norman, IEEE Trans. on Biomedical Engineering, Vol. 27, No. 1, pp. 105-114, Jan. 2000. (©2000 IEEE.)

 Analysis of Interaction Between ELF-LF Electric Fields and Human Bodies*

The interaction of extremely low frequency (ELF, 0-100 Hz) electromagnetic (EM) fields with the human body has become an increasingly important subject since potential health hazards due to the EM fields emitted by extremely high-voltage (EHV) power lines and ELF antenna systems became a public concern.

This subject has been extensively investigated experimentally or empirically by many workers [5]-[10]. However, all of the experiments were conducted on animals or scale models of man, and it is necessary to extrapolate these experimental data to provide data for human risk analysis. This is not an easy task if a reliable theoretical method for predicting the interaction of ELF fields with the human body is not available.

Theoretical studies on this subject have been conducted by a number of researchers, but they invariably used oversimplified body geometries or inaccurate methods. Shiau and Valentino [11] have used spheroidal models of man and their results from this idealized model may have little practical value. Spiegel [12], [13] used a more accurate block model and an electric field integral equation method, but his results disagree with experimental results mainly due to insufficient partition of the body model in the numerical calculation. Chiba et al. [14] used a finite-element method and a body of revolution geometry. Their results are still not accurate for a realistic human body. Kaune and McCreary [15] developed a numerical method on a cylindrical model of man. Since this model is over-simplified, the practical values of their results are questionable.

We have developed a numerical method which utilizes a realistic model of man with arbitrary shape and posture, and a realistic environmental condition such as assuming arbitrary grounding impedances between some parts of the body and ground [16]. Our method is developed on the basis of an integral equation for the induced surface-charge density, Ohm's law, and the

*The material in this section is based on "Quantification of interaction between ELF-LF electric fields and human bodies" by Kun-Mu Chen, Heuy-Ru Chuang and Chun-Ju Lin, which appeared in IEEE Transactions on Biomedical Engineering, Vol. BME 33, No. 8, pp.746-756, Aug. 1986 (©1986 IEEE.)

conservation of electric charge. The accuracy of our method has been checked with the exact solutions of a spherical and a spheroidal body. In addition, it has been verified by experimental results [9], [10] on the induced electric fields at the surface of the body, the short-circuit current, and the induced current density inside the body at 60 Hz. We have also found that it was possible to predict environmental results on the interaction of the human body with HF fields [17] with our method. This seems to extend the validity of our method to LF or even up to the HF range. It is also noted that our method is numerically quite efficient.

We will describe the theoretical development of our numerical method and reports some results on the induced electric fields at the body surface and inside the body, the induced current density inside the body, the short-circuit current, and the effects of the grounding impedance on the induced current in a homogeneous body of realistic shape.

GEOMETRY AND APPROXIMATIONS

Consider a geometry of a human body standing on the ground and being exposed to an electric field in the ELF-LF range, as shown in Fig. 22. The contacts between the feet, the hands, or other parts of the body and the ground are represented by the grounding impedances $Z_{Li}(i=1-k)$. The shape and posture of the body can be realistic and arbitrary. The impressed electric field, which is maintained by a power line or other ELF or LF sources, is assumed to be spatially uniform over the body and oscillating with an angular frequency of ω. The time dependence factor of $\exp(j\omega t)$ will be assumed.

We aim to determine the surface-charge density and the electric field induced on the body's surface as well as the current density, the electric field, and the SAR induced inside the body. Also, we aim to determine the effects of the grounding impedance on the induced current. To simplify the problem, the following approximations, which have been proved valid [7], will be adopted.

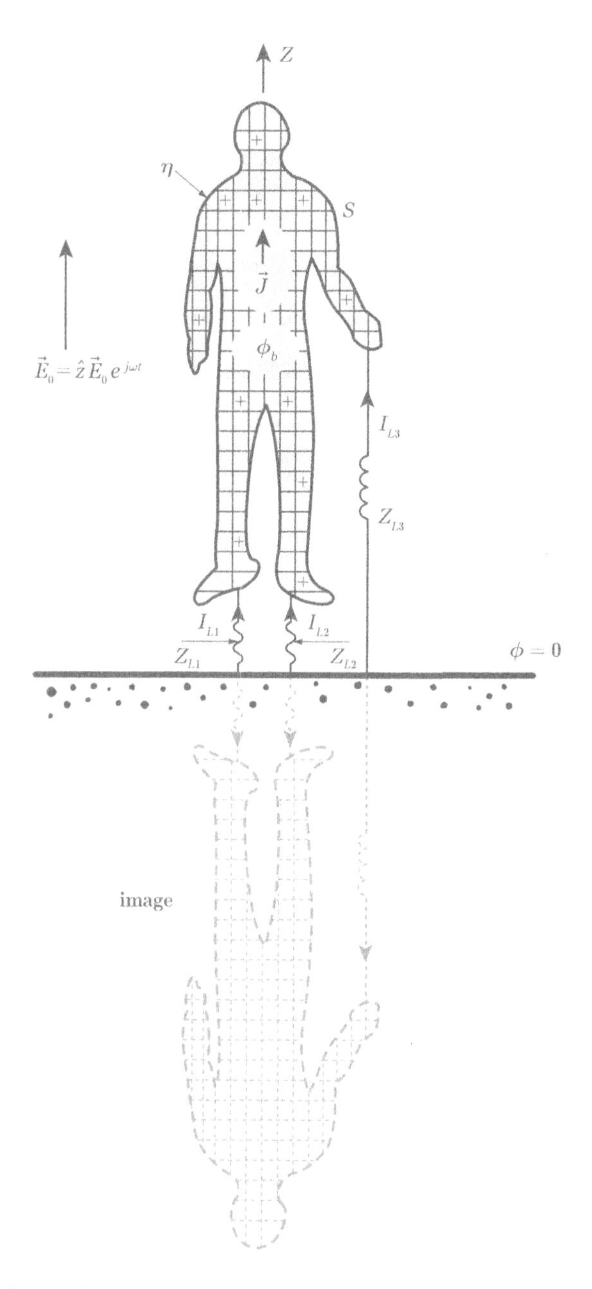

Fig. 22 A man standing on the ground is exposed to an electric field of ELF-LF range. The contacts between the feet and the left hand and the ground are represented by Z_{Li} $(i = 1-3)$.

1) The quasi-static approximation will be used since the body dimension is small compared to the wavelength of the impressed electric field.

2) The body will be assumed to be equipotential with an unknown potential ϕ_b.

3) The ground effect will be taken into account by the method of images.

4) The induced charge inside the body is assumed to be small compared to the induced charge on the body surface. Under the quasi-static approximation, the induced surface charge at the interface between media 1 and 2 is given by $\eta = \varepsilon_2(\sigma_1/\sigma_2 - \varepsilon_1/\varepsilon_2)(\hat{n}\cdot\vec{E}_1)$ where σ is the con-ductivity and ε is the permittivity. Thus, induced surface charge is proportional to the normal component of the $\vec{E}$ field in one region and also depends on the discontinuity of σ and ε of the media. $(\hat{n}\cdot\vec{E}_1)$ at the body's surface is many orders of magnitude higher than that inside of the body, and the discontinuities of σ and ε are greater at the body-air interface than at the interface of the different body tissues. Therefore, this assumption is valid.

5) Inside the body, the conduction current is assumed to be much greater than the capacitive current. This assumption is based on the fact that $\sigma >> \omega\varepsilon$ inside the body at the ELF-HF range.

It is noted that the effect due to the associated magnetic field will not be considered here because this effect is believed to be insignificant compared to that due to the electric field at this frequency range.

SURFACE-CHARGE INTEGRAL EQUATIONS

The first step in our approach is to determine the surface charge induced on the body surface by the impressed electric field. The induced surface charge density $\eta(\vec{r})$ and the body potential ϕ_b can be expressed as unknowns in a pair of integral equations as follows.

Physically, the body potential ϕ_b, which is spatially constant over the body, can be considered as the sum of the potential $\phi_S(\vec{r})$, which is maintained by the induced surface charge $\eta(\vec{r})$, and the potential $\phi_0(\vec{r})$, which is maintained by the impressed electric field. That is,

$$\phi_S(\vec{r})+\phi_0(\vec{r})=\phi_b \tag{4.10}$$

Using the quasi-static approximation and considering the ground image effect, $\phi_S(\vec{r})$ can be expressed as

$$\phi_S(\vec{r})=\frac{1}{4\pi\varepsilon_0}\int_{S+S_i}\eta(\vec{r})\frac{1}{|\vec{r}-\vec{r}'|}ds' \tag{4.11}$$

where s is the body surface and s_i is the surface of the body image, $\vec{r}$ is a field point on the body surface, and $\vec{r}'$ represents a source point on the body surface and the image surface. The potential $\phi_0(\vec{r})$ can be expressed in terms of the impressed electric field. For the geometry of Fig. 22, $\phi_0(\vec{r}')=-E_0 z$. The body potential ϕ_b is an unknown quantity and its value depends on the body geometry, the impressed electric field, and the grounding impedances Z_{Li}.

Equation (4.10) can be rewritten as

$$\frac{1}{4\pi\varepsilon_0}\int_{S+S_i}\eta(\vec{r}')\frac{1}{|\vec{r}-\vec{r}'|}ds'+\phi_0(\vec{r})=\phi_b \tag{4.12}$$

Equation (4.12) is an integral equation for the induced surface-charge density $\eta(\vec{r})$, and with the body potential ϕ_b as another unknown.

To determine $\eta(\vec{r})$ and ϕ_b, we need another equation. This second equation is obtained on the basis of Ohm's law and the conservation of electric charge. The total current flowing between the body and the ground is the sum of the currents flowing through the grounding impedances where Z_{Li}:

$$I=\phi_b\left[1/Z_{L1}+1/Z_{L2}+\ldots+1/Z_{Lk}\right]=\phi_b\sum_{i=1}^{k}1/Z_{Li} \tag{4.13}$$

On the other hand, the total current I (flowing from the ground to the body) can be expressed in terms of the total surface charge, based on the conservation of electric charge, as

$$I = j\omega \int_S \eta(\vec{r}')ds' \tag{4.14}$$

Combining Eq. (4.13) and Eq. (4.14), we have

$$\phi_b = \frac{j\omega}{\sum_{i=1}^{k} 1/Z_{Li}} \int_S \eta(\vec{r}')ds' \tag{4.15}$$

Eq. (4.15) is the desired second equation which has $\eta(\vec{r})$ and ϕ_b as the unknowns.

NUMERICAL SOLUTIONS—MOMENT METHOD

To solve Eq. (4.12) and Eq. (4.15) numerically for $\eta(\vec{r})$ and ϕ_b, the method of moments will be applied. The body surface S is partitioned into N subareas (patches) and the induced surface-charge density $\eta(\vec{r})$ on each subarea $\Delta s_n = (n = 1 - N)$ is assumed to be an unknown constant. Eq. (4.12) is then forced to be valid at the central points $\vec{r}_n\ (n = 1 - N)$ of the N subareas. In other words, the method of moments approach is applied using pulse basis functions and delta testing functions (point matching). When Eq. (4.12) is point matched at the center of the mth subarea $\vec{r}_m$, it can be expressed as

$$\sum_{i=1}^{k} \frac{1}{4\pi\varepsilon_0} \left[\int_{\Delta s_n} \frac{\eta_n ds'}{|\vec{r}_m - \vec{r}'|} - \int_{\Delta s_{ni}} \frac{\eta_n ds_i'}{|\vec{r}_m - \vec{r}_i'|} \right] + \phi_0(\vec{r}_m) = \phi_b \tag{4.16}$$

where η_n is the surface-charge density at the nth subarea Δs_n and $\vec{r}'$ is a source point within Δs_n.

The surface-charge density at the corresponding subarea Δs_{ni} of the body's image is $-\eta_n$ due to the image effect and $\vec{r}'$ is a source point within Δs_{ni}. Equation (4.16) can be rewritten as

$$\sum_{n=1}^{N} M_{mn}\eta_n + \phi_0(\vec{r}_m) = \phi_b \qquad (4.17)$$

where

$$M_{mn} = \frac{1}{4\pi\varepsilon_0}\left[\int_{\Delta s_n} \frac{ds'}{|\vec{r}_m - \vec{r}'|} - \int_{\Delta s_{ni}} \frac{ds_i'}{|\vec{r}_m - \vec{r}_i'|}\right] \qquad (4.18)$$

The integrals involved in Eq. (4.18) can be integrated numerically or analytically approximated when Δs_n is small. It s noted that when $n=m$ for M_{mm}, $\vec{r}_m$ is at the center of Δs_m and $\vec{r}'$ is within Δs_m; thus, $|\vec{r}_m - \vec{r}'|$ will vanish in the first integral of Eq. (4.18). However, this singularity is removable through the integration and it causes no difficulty.

Equation (4.17) can be used to generate *N* simultaneous equations when *m* is varied from 1 to N, that is, when Eq. (4.16) is point matched at the central points of the *N* subareas. This set of *N* simultaneous equations can be expressed in a matrix form as follows:

$$\begin{bmatrix} M_{11} & M_{12} & \cdots & M_{1N} & -1 \\ M_{21} & M_{22} & \cdots & M_{2N} & -1 \\ & & \cdots & & -1 \\ M_{N1} & M_{N2} & \cdots & M_{NN} & -1 \end{bmatrix} \begin{bmatrix} \eta_1 \\ \eta_2 \\ \vdots \\ \eta_N \\ \phi_b \end{bmatrix} = - \begin{bmatrix} \phi_{01} \\ \phi_{02} \\ \vdots \\ \phi_{0N} \end{bmatrix} \qquad (4.19)$$

where $\phi_{0n} = \phi_0(\vec{r}_n)$. Notice that Eq. (4.19) represents a set of equations that is of $N \times (N+1)$ order. In order to solve $N+1$ unknowns $(\eta_1, \eta_2, \cdots, \eta_N$ and $\phi_b)$, one more equation between η_n and ϕ_b is needed. This equation is provided by Eq. (4.15) as follows:

$$\sum_{n=1}^{N} \eta_n \Delta s_n = \frac{1}{j\omega}\left[\sum_{i=1}^{k} 1/Z_{Li}\right]\phi_b \qquad (4.20)$$

When Eq. (4.20) is combined with Eq. (4.19), we have

$$\begin{bmatrix} M_{11} & M_{12} & \cdots & M_{1N} & -1 \\ M_{21} & M_{22} & \cdots & M_{2N} & -1 \\ & & \cdots & & -1 \\ M_{N1} & M_{N2} & \cdots & M_{NN} & -1 \\ \Delta s_1 & \Delta s_2 & & \Delta s_N & \frac{j}{\omega}\sum_{i=1}^{k} 1/Z_{Li} \end{bmatrix} \begin{bmatrix} \eta_1 \\ \eta_2 \\ \vdots \\ \eta_N \\ \phi_b \end{bmatrix} = - \begin{bmatrix} \phi_{01} \\ \phi_{02} \\ \vdots \\ \phi_{0N} \\ 0 \end{bmatrix} \tag{4.21}$$

Equation (4.21) represents an $(N+1)\times(N+1)$ matrix equation, and $(N+1)$ unknowns, $\{\eta_n\}$ and ϕ_b, can be easily determined by a matrix inversion or other appropriate methods.

There are two special cases of interest: 1) the case when the body is shorted to the ground, $Z_{Li}=0$ for any i, and 2) the case when the body is isolated from the ground, $Z_{Li}=\infty$ for all i. For the short-circuit case, the body potential ϕ_b will be zero; therefore, Eq. (4.21) is reduced to an $N\times N$ matrix equation, with the last column and the last row of the matrix in Eq. (4.21) removed. The unknown surface-charge density $\{\eta_n\}$ can then be determined accordingly. For the isolated case, the last element of the matrix of Eq. (4.21), $\frac{j}{\omega}\sum_{i=1}^{k}\frac{1}{Z_{Li}}$ becomes zero. This implies that

$$\sum_{n=1}^{N} \eta_n \Delta s_n = 0 \tag{4.22}$$

that is, the total net charge on the body is zero. For this case, $\{\eta_n\}$ and ϕ_b are determined from Eq. (4.21) with the last element of the matrix set equal to zero.

After the induced surface-charge density η is determined, the induced electric field at the body surface is simply obtained from

$$E_s = \eta / \varepsilon_0 \tag{4.23}$$

assuming that the induced electric field is totally perpendicular to the body surface.

The electric field enhancement factor is defined as the ratio of the induced electric field at the surface to the impressed electric field E_S/E_0. This value can easily exceed 10 at the head or the tip of a stretched arm and hand. Thus, when a man is exposed to the electric field of an EHV power line, the induced electric field at some points of the body can be extremely large since E_0 is already a very high value in this case.

INDUCED CURRENT INDISE THE BODY

After the induced surface-charge density at any point on the body surface is determined, the induced current density inside the body can be determined on the basis of the conservation of electric charge and Maxwell's equations.

The first quantity to be determined is the total sectional current at any cross section of the body. Referring to Fig. 23, we assume that the positive sectional current at any cross section of the body is directed towards the head. We will consider, for example, three sectional currents: the sectional current at the chest I_1, the sectional current at the lower abdomen I_2, and the sectional current at the right arm I_3.

If we integrate the equation of the conservation of electric charge $\nabla \cdot \vec{J} + j\omega\rho = 0$ over the volume V_1 which includes the upper body above the chest cross section (see Fig. 23), we have

$$I_1 = -\int_{S_{c1}} \left(\hat{n} \cdot \vec{J}_1\right) ds = j\omega \int_{S_1} \eta \, ds \tag{4.24}$$

where $\hat{n}$ is the unit vector pointing outward from V_1, S_{c1} is the cross-sectional area at position shown in Fig. 23, S_1 is the body surface above the chest cross section S_{c1}, and $\vec{J}_1$ is the current density at S_{c1}.

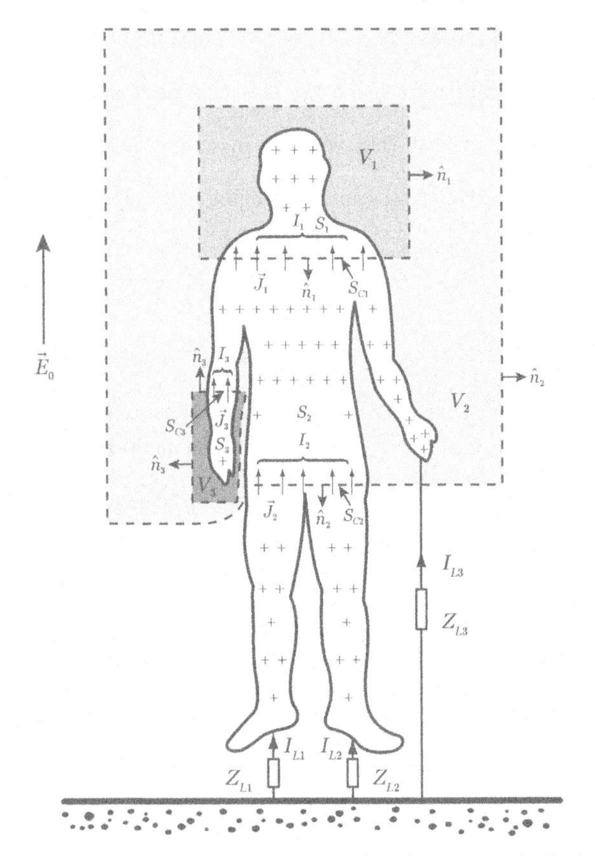

Fig. 23 Geometry for calculating the induced current in the body.

A similar integral over the volume V_2, which includes the portion of the body above the lower abdomen and has a boundary surface cutting through the lower abdomen section S_{c2} and a grounding impedance Z_{L3} connecting the left hand to ground, will lead to

$$\int_{S_{c2}} \left(\hat{n}_2 \cdot \vec{J}_2\right) ds - I_{L3} = -j\omega \int_{S_2} \eta \, ds$$

or

$$I_2 + I_{L3} = j\omega \int_{S_2} \eta \, ds \tag{4.25}$$

where $\hat{n}_2$ is the unit vector pointing outward form V_2, $\vec{J}_2$ is the current density at S_{c2}, S_2 is the body surface enclosing by V_2, and I_{L3} is the current flowing from the ground to the left hand through the grounding impedance Z_{L3}. I_{L3} is easily determined from ϕ_b/Z_{L3}.

Another similar integral over the volume V_3 which contains the right arm and hand leads to

$$I_3 = \int_{S_{c3}} \left(\hat{n}_3 \cdot \vec{J}_3\right) ds = -j\omega \int_{S_3} \eta \, ds \tag{4.26}$$

where $\hat{n}_3$ is the unit vector pointing outward form V_3, S_{c3} is the cross-sectional area as designated in Fig. 23, $\vec{J}_3$ is the current density at S_{c3}, and S_3 is the surface of the right arm and right hand enclosed by V_3. It is noted that $\hat{n}_3$ and $\vec{J}_3$ are in the same direction; therefore, the expression of I_3 has a negative sign in Eq. (4.26). This negative sign will lead to the phenomenon that the current in the arm flows in the opposite direction to that flowing in other parts of the body.

One of the most important quantities concerning the body current is the short-circuit current I_{sc}, which is defined as the current flowing between the feet and the ground when the grounding impedances (Z_{L1} and Z_{L2}) between the feet and ground are zero. The other grounding impedances (e.g., Z_{L3}) between other parts of the body and ground are infinity (open circuit). I_{sc} can be easily obtained by the same approach as above, and is given by

$$I_{sc} = j\omega \int_s \eta \, ds \tag{4.27}$$

where S includes the total body surface and η is the induced surface-charge density at the body surface under the condition that the body's potential ϕ_b is zero.

After the determination of the sectional current, the volume density of the induced current inside the body $\vec{J}$ can be determined from Maxwell's equation: $\nabla \times \vec{H} = (\sigma + j\omega\varepsilon)\vec{E}$ or $\nabla \cdot \left[(\sigma + j\omega\varepsilon)\vec{E}\right] = 0$. For the ELF-LF range, $\sigma >> \omega\varepsilon$ inside the body; therefore,

$$\nabla \cdot \left(\sigma \vec{E}\right) = \nabla \cdot \vec{J} \doteq 0 \tag{4.28}$$

Equation (4.28) can be used to predict the distribution of $\vec{J}$ inside the body with the prior knowledge of the sectional current I at any cross section of the body.

Assume that $\vec{J}$ at any cross section of the body has only two components: a longitudinal component $\vec{J}_l$ and a radial component $\vec{J}_r$. This approximation assumes a cylindrical geometry for the body cross section and also ignores the circumferential component of $\vec{J}$.

The longitudinal component $\vec{J}_l$, can be approximately obtained as

$$\vec{J}_l = I / s_c \tag{4.29}$$

where I is the already determined total sectional current and s_c, is the cross-sectional area of the body at the position where Eq. (4.29) is applied. The calculation of $\vec{J}_l$ is valid for a homogeneous body, but it may also be a fair approximation for a heterogeneous body, based on the finding by Spiegel [13] using the electric field integral equation method that the induced current density $\vec{J}$ is rather independent of the electric parameters of the body at the ELF range.

Now that $\vec{J}_l$ is determined at any cross section of the body, the radial component $\vec{J}_r$ can be derived from Eq. (4.28), using a cylindrical geometry, as follows.

From

$$\nabla \cdot \vec{J} = \frac{1}{r}\frac{\partial}{\partial r}(rJ_r) + \frac{\partial}{\partial l}J_l = 0,$$

we have

$$\frac{\partial}{\partial r}(rJ_r) = -r\left[\frac{\partial J_l}{\partial l}\right]$$

After integrating both sides, it gives

$$J_r = -\frac{r}{2}\left[\frac{\partial J_l}{\partial l}\right] \tag{4.30}$$

where r is the radical distance from the center of the cross section and $[\partial J_l / \partial l]$ is the rate of change of $\vec{J}_l$, in the longitudinal direction. Since $\vec{J}_l$, is known at any cross section, the value of $[\partial J_l / \partial l]$ can be estimated easily. Equation (4.30) indicates that $\vec{J}_r$ is zero at the center of the body and linearly increases toward the body surface. The direction of $\vec{J}_r$ is dictated by the sign of $[\partial J_l / \partial l]$.

After $\vec{J}$ is determined, the electric field induced inside the body is determined from

$$\vec{E} = \vec{J} / \sigma \tag{4.31}$$

and the SAR value is calculated via

$$\text{SAR} = \frac{1}{2\sigma} |\vec{J}|^2 / \rho \ \text{W/kg} \tag{4.32}$$

where ρ is the volume density of mass in kg/m^3.

NUMERICAL RESULTS AND COMPARISON TO EXPERIMENTS

We have generated many useful numerical results on the induced electric fields at the body surface and inside the body, the induced current, the SAR, and effects of the grounding impedance on the short-circuit current. Since these results are quite extensive, only a few numerical results will be compared to the experimental results to validate the accuracy of the present method.

The first check of our method was made by comparing the numerical results of the induced electric field on an isolated conducting sphere to its exact solution. The agreement was found to be within 1 percent. We then verified the accuracy of our method by comparing the numerical

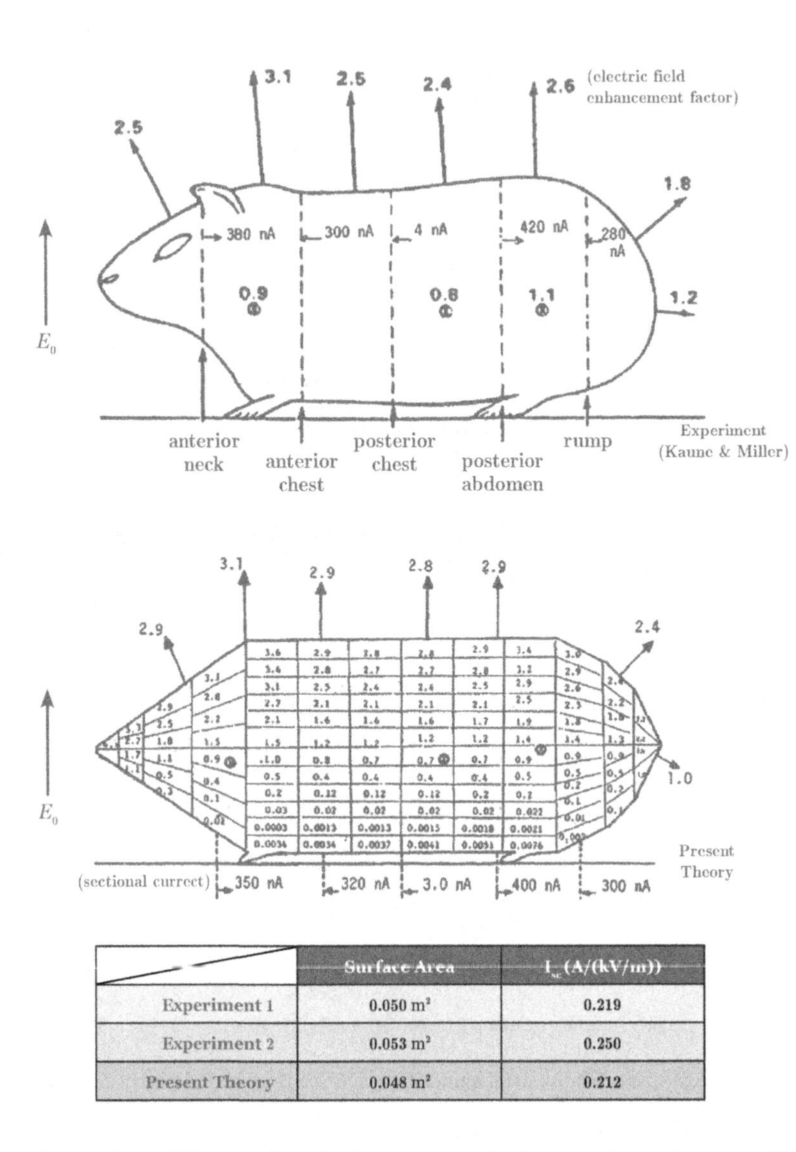

	Surface Area	I_{sc} (A/(kV/m))
Experiment 1	0.050 m^2	0.219
Experiment 2	0.053 m^2	0.250
Present Theory	0.048 m^2	0.212

Fig. 24 Comparison of theoretical results by present method to experimental results of Kaune and Miller on electric field enhancement factor, sectional current, and short-circuit currents for grounded guinea pig exposed to 10 kV/m, 60 Hz electric field.

results on the electric field enhancement factor, the sectional current, and the short-circuit current induced by a 60 Hz electric field of 10 kV/m in a theoretical model of a guinea pig which simulated the actual animal used by Kaune and Miller [9] to their experimental results. Fig. 24 shows the comparison between the experimental and numerical results. The upper part of Fig. 24 shows the electric field enhancement factors measured at various locations on the surface and the sectional currents at five cross sections of a guinea pig. The lower part shows the electric field enhancement factors at various points and five sectional currents calculated for the theoretical model of the guinea pig. Good agreement is obvious when the corresponding experimental and numerical values of these quantities shown in these two figures are compared. The table at the bottom of Fig. 24 indicates the comparison of the measured short-circuit current and the calculated values. The measured short-circuit currents for two guinea pigs were 0.219 and 0.225 μA/(kV/m) and our calculated result was 0.212 μA/(kV/m). The results shown in Fig. 24 demonstrate the accuracy of our method. It is noted that in the numerical calculation, the body surface of the guinea pig was partitioned into 228 patches. This leads to 114 unknowns when a half-body symmetry was applied.

Our method was also employed to compute the current density induced by a 60 Hz electric field of 10 kV/m in a phantom model of man, 45 cm in height, used by Kaune and Forsythe [10]. Fig. 25 depicts the comparison of the experimental and numerical results of the induced current density. In the numerical calculation, the body surface was partitioned into 472 patches, leading to 118 unknowns with a quarter-body symmetry. The right figure shows the measured current densities at various points inside the body. The induced current density is mainly longitudinal (or vertical) with a small radial (or horizontal) component as shown.

Notice that the direction of the radial component is outward in the chest region, but is inward at the neck and the abdominal region. The most interesting observation is that the induced current in the arm is directed downward or in the opposite direction to that flowing in other parts of the body. The numerical results for the induced current density in the theoretical

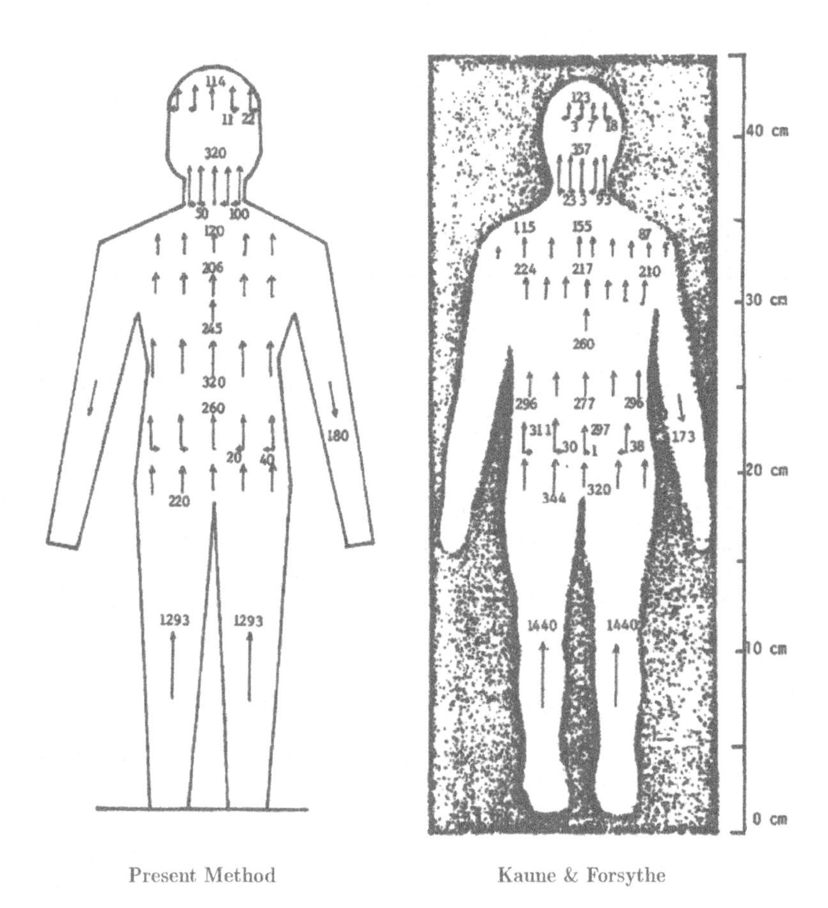

Fig. 25 Comparison of theoretical results by the present method to experimental results of Kaune and Forsythe on vertical and horizontal current densities for a grounded human model exposed to 10 kV/m, 60 Hz electric field. Induced current densities are given in units of $n\mathrm{A/cm^2}$.

model which approximates the experimental model are shown in the left figure. The computed current densities, in amplitude and direction, at various locations inside the body agree very well with the measured values.

The agreement between experiment and theory is within 5-10 percent on the amplitude. It is noted that the theory correctly predicted the reversed direction of the induced current in the arm and the directions of the radial components of the currents at different parts of the body. The results of Fig. 25 give a positive verification of the accuracy of our method.

It is appropriate to give a few more selected numerical results on the interaction of a 60 Hz electric field with a human body. Fig. 26 depicts the calculated electric field enhancement factors at the surface of a man with a height of 180 cm and a weight of 68.2 kg standing upright and in direct contact (short-circuited case) with the ground, and who is exposed to a 60 Hz electric field of 1 kV/m. The enhancement factor can vary from 0.1 to about 20 over the body surface. Also shown in Fig. 26 is the calculated short-circuit current of 18.0 μA. This value is very close to 17.5 μA, which is calculated with an empirical formula of $I_{sc} = 5.4 \cdot 10^{-9} \cdot H^2 \cdot E \cdot f / 60$ used by Chiba et al. [14]. In the numerical calculation, the body's surface was partitioned into 424 patches leading to 106 unknowns with a quarter-body symmetry.

Fig. 27 shows the calculated electric field enhancement factors and the short-circuit current in the same man with stretched arms induced by the same electric field as the case of Fig. 26. It is observed that the electric field enhancement factor can be very high at the tip of the hand due to its sharp geometry. Also, it is noted that when the arms are stretched, the short-circuit current is increased to 23.3 μA as predicted by our method. This value is quite different from the value of 17.5 μA if the same formula used by Chiba et al. [14] is used. This indicates the phenomenon that the induced electric field at the body surface and the short-circuit current, and consequently the induced current inside the body, are strongly dependent on the body geometry and position. Even though the phenomena involved are rather complicated, this method is capable of predicting them.

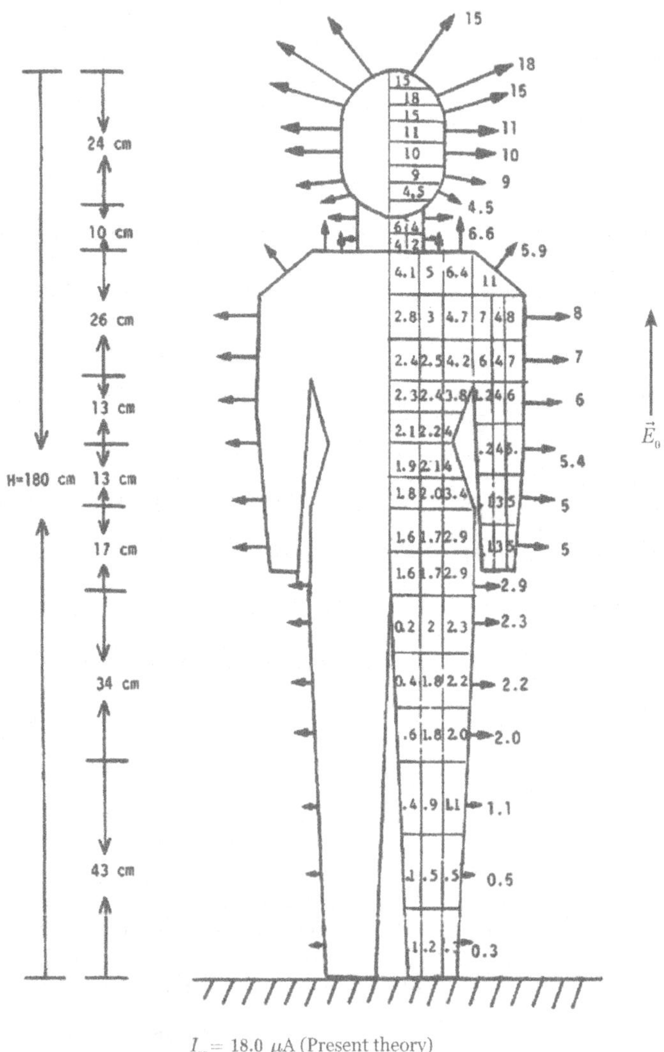

I_{sc} = 18.0 μA (Present theory)

I_{sc} = 17.5 μA (Based On $I_{sc} = 5.4 \cdot 10^{-9} \cdot H^2 \cdot E \cdot f/60$ see Chiba et al.)

Fig. 26 Theoretical results on electric field enhancement factors and short-circuit current for a realistic model of man standing on the ground being exposed to 1kV/m, 60 Hz electric field.

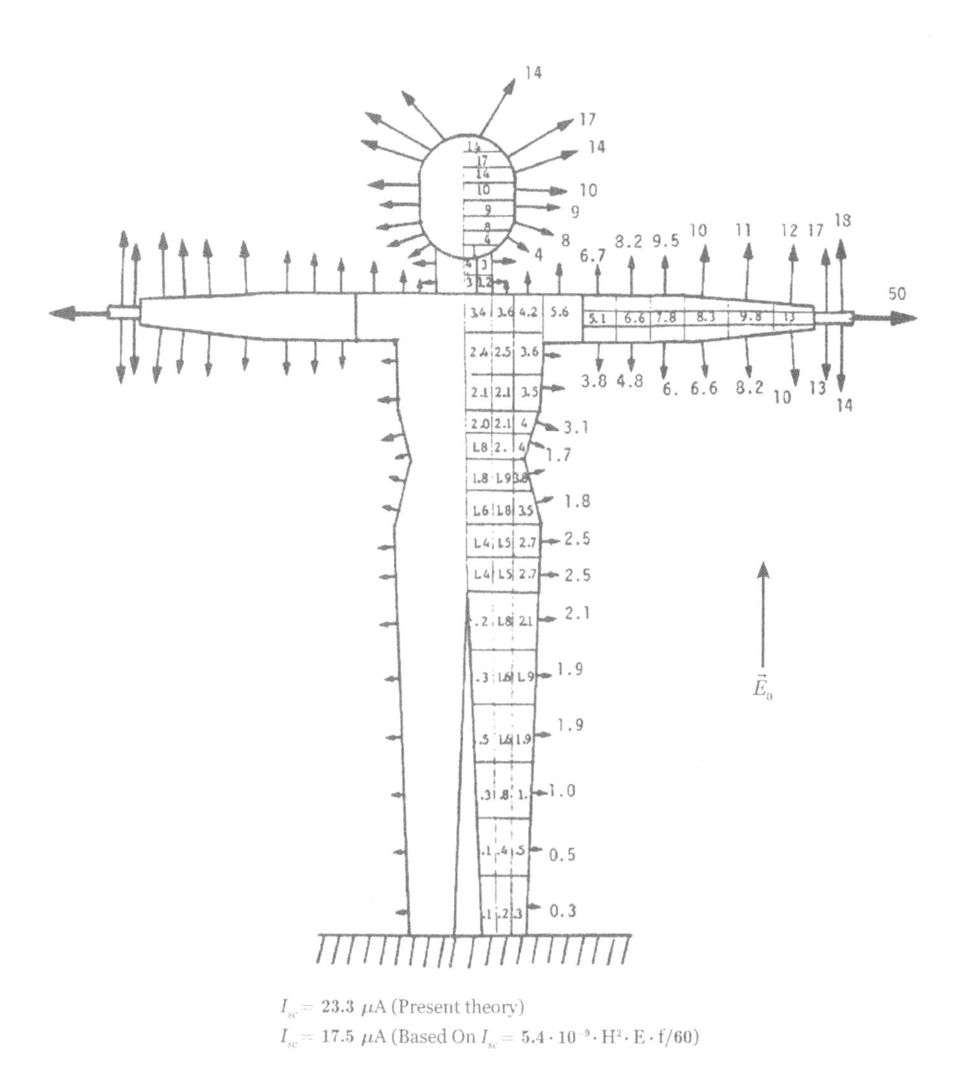

Fig. 27 Theoretical results on electric field enhancement factors and short-circuit current for a realistic model of man with hands stretching horizontally and standing on the ground plane being exposed to 1 kV/m, 60 Hz electric field.

One more bit of interesting information on the effects of the grounding impedance on the short-circuit current of a man exposed to 60 Hz electric field of 1 kV/m is given in Fig. 28. The grounding impedance is assumed to be resistive, capacitive, or inductive. The short-circuit current is found to remain practically unchanged, maintaining a value of about 18 μA/(kV/m) when the grounding impedance is varied from 0 to 10 MΩ. Only after the grounding impedance exceeds the value of 10 MΩ does the short-circuit current start to fall for a resistive or a capacitive grounding impedance. For an inductive grounding impedance, there is a possible resonance

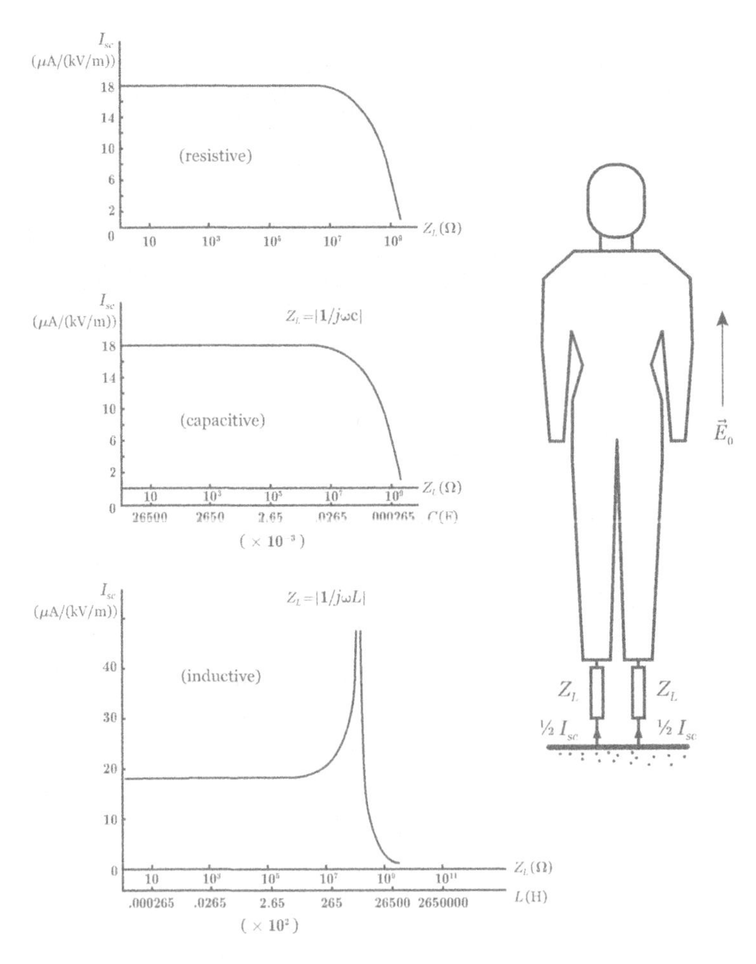

Fig. 28 Short-circuit current as a function of grounding impedance.

phenomenon when the impedance has a value on the order of 100 MΩ, corresponding to an inductance of 3×10^5 H. This implies that in a very unlikely case when the grounding impedance is an extremely large inductance of about 3×10^5 H, there may be a very large current induced in the body.

The numerical examples given so far are all computed for the case of 60 Hz. However, our method has also been employed to predict the short-circuit currents induced by HF electric fields in a human body. Table I shows a comparison of experimental results on the short-circuited current, measured by various workers [17] at frequencies between 60 Hz and 27 MHz, and the corresponding numerical results generated by our method. A satisfactory agreement was obtained between theory and experiment up to about 30 MHz. This indicates that our method may be applicable up to the HF range or at least up to the LF range.

Table I

Comparison of experimental and empirical results on short-circuit current and theoretical results by present method

f_{MHz}	I_{sc} in (mA/(V/m))		
	$h_m = 1.75$ m	$h_m = 1.75$ m	$h_m = 1.80$ m
	Eq.(*)	Measured	Present Theory (SCIE)
Subjects Barefoot			
0.63	0.208	0.210	0.189
0.70	0.232	0.280	0.212
1.51	0.499	0.391	0.453
27.405	9.060	9.330	8.230
Subjects Barefoot, Both Arms Raised			
0.70		0.384	0.272
1.51		0.555	0.586
27.405		9.85	10.60
Comparison Data			
0.72 [13]	0.238	0.277	0.216
0.92	0.304	0.316	0.276
1.145	0.379	0.366	0.344
1.35	0.447	0.405	0.405
1.47	0.486	0.56	0.441
0.146 [14]	0.048	0.035	0.044
27.0 [15]	8.93	8.4	8.13

* $I_{sc} = 0.108 \cdot h_m^2 \cdot E_0 \cdot f_{MHz}$[mA/(V/m)].

REFERENCES

[1] Chen, K.M., Misra, D., Wang, H., Chuang, H.R. and Postow, E., "An X-band Microwave Life-Detection System," *IEEE Trans. On Biomedical Engineering*, Vol. 33, pp. 697-701, 1986.

[2] Chen, Kun-Mu, "Active remote physiological monitoring using Microwaves," *Medical monitoring in the home and work environment*, Edited by Laughton E. Miles and Roger J. Broughton, Raven Press, New York (1990).

[3] Harrington, R.F. (1961), *Time Harmonic Electromagnetic Fields*, McGraw-Hill Book Co., New York.

[4] Chen Kun-Mu, Yong Huang, Jianping Zhang, and Adam Norman, "Microwave life-detection systems for searching human subjects under earthquake rubble or behind barrier," *IEEE Trans. on Biomedical Engineering*, Vol. 27, No.1, pp. 105-114, Jan. 2000.

[5] D. W. Deno, "Currents induced in the human body by high voltage transmission line electric field-Measurement and calculation of distribution and dose," *IEEE Trans. Power App. Syst.*, vol. PAS-96, pp. 1517-1527, Sept./Oct. 1977.

[6] A. W. Guy, M. D. Webb, and C. C. Sorensen, "Determination of power absorption in man exposed to high frequency electromagnetic fields by thermographic measurements on scale models," *IEEE Trans. Biomed. Eng.*, vol. BME-23, pp. 361-371, 1976.

[7] W. T. Kaune and M. F. Gillis, "General properties of the interaction between animals and ELF electric fields," *Bioelectromagn.*, vol. 2, pp. 1-11, 1980.

[8] W. T. Kaune and R. D. Phillips, "Comparison of the coupling of grounded humans, swine and rats to vertical, 60-Hz electric fields," *Bioelectromagn.*, vol. 1, pp. 117-129, 1980.

[9] W. T. Kaune and M. C. Miller, "Short-circuit currents, surface electric fields, and axial current densities for guinea pigs exposed to ELF electric fields," *Bioelectromagn.*, vol. 5, pp. 361-364, 1984.

[10] W. T. Kaune and W. C. Forsythe, "Current densities measured in human models exposed to 60 Hz electric fields," *Bioelectromagn.*, vol. 6, pp. 13-32, 1985.

[11] Y. Shiau and A. R. Valentino, "ELF electric field coupling to dielectric spheroidal models of biological objects," *IEEE Trans. Biomed. Eng.*, vol. BME-28, pp. 429-437, June 1981.

[12] R. J. Spiegel, "High-voltage electric field coupling to humans using moment method techniques," *IEEE Trans. Biomed. Eng.*, vol. BME- 24, pp. 466-472, Sept. 1977.

[13] –, "Numerical determination of induced currents in humans and baboons exposed to 60-Hz electric fields," *IEEE Trans. Electromagn. Compat.*, vol. EMC-23, pp. 382-390, Nov. 1981.

[14] A. Chiba, K. Isaka, and M. Kitagawa, "Application of finite element method to analysis of induced current densities inside human model exposed to 60-Hz electric field," *IEEE Trans. Power App. Syst.*, vol. PAS-103, pp. 1895-1902, July 1984.

[15] W. T. Kuane and F. A. McCreary, "Numerical calculation and measurement of 60-Hz current densities induced in an upright grounded cylinder," *Bioelectromagn.*, vol. 6, pp. 209-220, 1985.

[16] Kun-Mu Chen, Huey-Ru Chuang and Chun-Ju Lin, "Quantification of interaction between ELF-LF electric fields and human bodies," *IEEE Trans. Biomed. Eng.*, vol. BME-33, No.8, pp 746-756, Aug. 1986.

[17] O. P. Gandhi, I. Chatterjee, D. Wu, and Y. G. Gu, "Likelihood of high rates of energy deposition in the human legs at ANSI recommended 3-30 MHz RF safety levels," *Proc. IEEE*, vol. 73, pp. 1145- 1147, June 1985.

Chapter 5
Radar Target Identification with Extinction-pulse (E-pulse) Method

In this chapter we will study a new method on the radar target identification using an ultra-wide band radar and the Extinction pulse (E-pulse) scheme.

A conventional radar can detect the location and estimate the size of the target only. Some existing target identification methods include the cooperative method such as Friend or Foe device. This method is not a fool-proof method. It is desirable to construct a non-cooperative method which does not need a cooperation of the target under the interrogation by a radar signal.

One of the difficulties involved in a non-cooperative target identification is the fact that when a target is illuminated by an incident radar signal, its radar return is strongly aspect-dependent. The magnitude and shape of the radar return vary for different aspects of the incident radar signal. However, the natural-frequencies of the late-time part of the radar return are aspect-independent. We will make use of these unique characteristics to develop the so called Extinction pulse (E-pulse) method to provide an aspect-independent target identification scheme.

First, the time-domain analysis of the E-pulse method will be presented. Then, the frequency-domain synthesis of the E-pulse method will be studied. Lastly, the E-pulse method will be applied to discriminate the early-time radar return.

 Radar Target Identification by E-pulse Method — Time-Domain Analysis

This non-cooperative, aspect-independent radar target Identification scheme was developed by researchers at Michigan State University in 1980-90 [1]-[6]. The basic principle is to illuminate the radar target with a short EM pulse (or ultra-wide band signal) and use the late-time response of the received radar return to identify the target.

When a radar target is illuminated by a short pulse, the scattered radar return consists of an early-time response and a late-time response. The early-time response is a collection of specular reflections from the structural discontinuities of the target and it is strongly dependent on the aspect angle of the incident pulse. The late-time response is created by the free-oscillation of the induced current on the target. Its shape is dependent on the aspect angle of the incident pulse but the contents of the natural frequencies of the late-time response are independent of the aspect angle of the incident pulse.

We have found that we can synthesize a discriminant signal based on the target's natural frequencies, called the extinction-pulse or the E-pulse. The E-pulse can be used to convolve with the late-time response of the target's radar return to produce a zero response or a single-mode response in the late-time period. Since the E-pulse is synthesized using the natural frequencies of the target, it is unique with a particular target and it is aspect-independent. It is noted that when the E-pulse of a particular target is convolved with the radar return of a wrong target, the convolved output will be significantly different from the expected zero or a single-mode response. Thus an aspect-independent target identification can be achieved.

SYNTHESIS OF DISCRIMINANT SIGNALS BASED ON NATURAL FREQUENCIES

When a target is illuminated by an incident pulse as shown in the following figure, the received radar return consists of an early-time response and a late-time response.

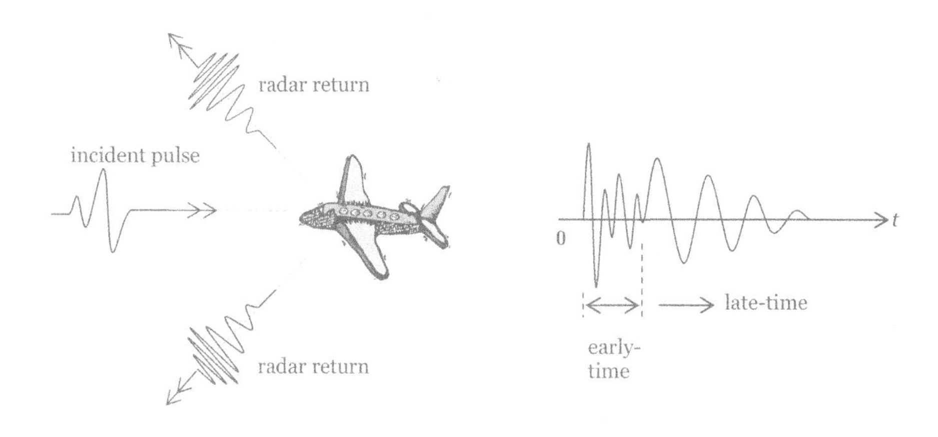

The late-time radar return of a target is the sum of natural modes and can be expressed as

$$h(t) = \sum_{n=1}^{N} a_n(\theta,\phi) e^{\sigma_n t} \cos(\omega_n t + \phi_n(\theta,\phi)) \tag{5.1}$$

where σ_n = damping coefficient of the nth natural mode

ω_n = angular frequency of the nth natural mode

σ_n, ω_n are independent of the aspect-angle (θ, ϕ)

$a_n(\theta, \phi)$ = amplitude of the nth natural mode

$\phi_n(\theta, \phi)$ = phase angle of the nth natural mode

N = total number of natural modes which are excited by an incident radar pulse (number of modes excited are determined by the bandwidth of the incident pulse)

a_n, ϕ_n are aspect-dependent

Analogies in mechanical systems on the independence of the natural frequencies on the aspect angle of the incident pulse are shown bellows. A turning fork can be hit at various angles while the natural frequencies of the produced sound stay constant. A violin string can be pecked at various locations but it produces a tune consists of the same natural frequencies.

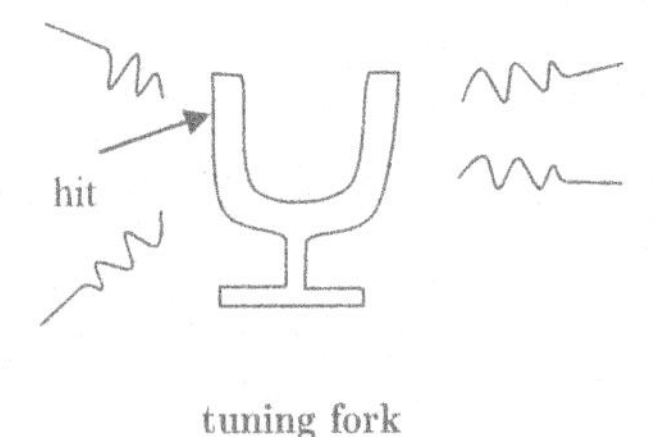

tuning fork

violin string

natural frequencies are independent of how they are striken.

We now aim to synthesize an extraction signal of duration T_e which can be convolved with the radar return $h(t)$ to produce a single-mode or zero mode output in the late-time period $(t > T_e)$:

$$E^o(t) = \int_0^{T_e} E^e(t')h(t-t')dt' \qquad \text{for } t > T_e \tag{5.2}$$

where

$E^0(t) =$ convolved output signal

$E^e(t) =$ extraction signal to be synthesized

Substituting Eq. (5.1) into Eq. (5.2),

$$E^o(t) = \int_0^{T_e} E^e(t') \sum_{n=1}^{N} a_n e^{\sigma_n (t-t')} \cos[\omega_n (t-t') + \phi_n] dt'$$

$$= \int_0^{T_e} E^e(t') \sum_{n=1}^{N} a_n e^{\sigma_n t} e^{-\alpha_n t'} [\cos(\omega_n t + \phi_n) \cos\omega_n t' + \sin(\omega_n t + \phi_n) \sin\omega_n t'] dt'$$

$$= \sum_{n=1}^{N} a_n e^{\sigma_n t} \Bigg[\cos(\omega_n t + \phi_n) \underbrace{\int_0^{T_e} E^e(t') e^{-\alpha_n t'} \cos\omega_n t' dt'}_{A_n}$$

$$+ \sin(\omega_n t + \phi_n) \underbrace{\int_0^{T_e} E^e(t') e^{-\alpha_n t'} \sin\omega_n t' dt'}_{B_n} \Bigg]$$

$$= \sum_{n=1}^{N} a_n e^{\sigma_n t} [A_n \cos(\omega_n t + \phi_n) + B_n \sin(\omega_n t + \phi_n)] \tag{5.3}$$

for $t > T_e$

where

$$A_n = \int_0^{T_e} E^e(t') e^{-\alpha_n t'} \cos\omega_n t' dt' \tag{5.4}$$

$$B_n = \int_0^{T_e} E^e(t') e^{-\alpha_n t'} \sin\omega_n t' dt' \tag{5.5}$$

A_n and B_n determine the amplitude of the natural modes of the convolved output, and are independent of the aspect angle (θ and ϕ). This makes it possible to synthesize the aspect-independent $E^e(t)$. It is noted that A_n and B_n are numerically stable because they are finite integrals over a short period of T_e even though there is a time growing factor of $e^{-\sigma_n t}$ in them.

If $E^e(t)$ is synthesize in such a way that

$$A_1 = 1 \text{ and all other } A_n \text{ and } B_n = 0$$

then the convolved output will be

$$E^0(t) = a_1(\theta,\phi) e^{\sigma_1 t} \cos(\omega_1 t + \phi_1(\theta,\phi)) \qquad \text{for } t > T_e$$

If $E^e(t)$ is synthesized in a way that

$$B_3 = 1, \text{ and all other } A_n \text{ and } B_n = 0$$

then the convolved output will be

$$E^0(t) = a_3(\theta,\phi)e^{\sigma_3 t}\sin(\omega_3 t + \phi_3(\theta,\phi)) \qquad \text{for } t > T_e$$

If $E^e(t)$ is synthesized in such a way that

$$\text{all } A_n \text{ and } B_n = 0,$$

then the zero-mode output is obtained :

$$E^0(t) \qquad \text{for } t > T_e$$

These synthesized extraction signals are called simple-mode extraction and zero-mode extraction signals.

Next, how to synthesize $E^e(t)$? Construct $E^e(t)$ with a set of basis functions $f_m(t)$ as

$$E^e(t) = \sum_{m=1}^{2N} d_m f_m(t) \tag{5.6}$$

where $\{f_m(t)\}$ can be pulse functions or other functions, and $\{d_m\}$ are unknown coefficients to be determined. We need $2N$ of $f_m(t)$ because there are $2N$ of A_n and B_n to be specified.

$$\begin{aligned} A_n &= \int_0^{T_e} E^e(t')e^{-\sigma_n t'}\cos\omega_n t' dt' \\ &= \int_0^{T_e} \sum_{m=1}^{2N} d_m f_m(t')e^{-\sigma_n t'}\cos\omega_n t' dt' \\ &= \sum_{m=1}^{2N} d_m \underbrace{\int_0^{T_e} f_m(t')e^{-\sigma_n t'}\cos\omega_n t' dt'}_{M_{nm}^c} \end{aligned}$$

$$A_n = \sum_{m=1}^{2N} d_m M_{nm}^c \tag{5.7}$$

while

$$B_n = \int_0^{T_e} E^e(t')e^{-\sigma_n t'} \sin\omega_n t' dt'$$
$$= \int_0^{T_e} \sum_{m=1}^{2N} d_m f_m(t') e^{-\sigma_n t'} \sin\omega_n t' dt'$$
$$= \sum_{m=1}^{2N} d_m \underbrace{\int_0^{T_e} f_m(t') e^{-\sigma_n t'} \sin\omega_n t' dt'}_{M_{nm}^s}$$

$$B_n = \sum_{m=1}^{2N} d_m M_{nm}^s \tag{5.8}$$

Equations (5.7) and (5.8) can be expressed in a matrix form as

$$\begin{bmatrix} A_1 \\ A_2 \\ \vdots \\ B_1 \\ \vdots \\ B_N \end{bmatrix} = \begin{bmatrix} M_{11}^c & M_{12}^c & \cdots & M_{12N}^c \\ M_{21}^c & M_{22}^c & \cdots & M_{22N}^c \\ \vdots & \vdots & \ddots & \vdots \\ M_{11}^s & M_{12}^s & \cdots & M_{12N}^s \\ \vdots & \vdots & \ddots & \vdots \\ M_{N1}^s & M_{N2}^s & \cdots & M_{N2N}^s \end{bmatrix} \begin{bmatrix} d_1 \\ d_2 \\ \vdots \\ d_{N+1} \\ \vdots \\ d_{2N} \end{bmatrix} \tag{5.9}$$

where $n = 1, 2, \cdots, N \quad m = 1, 2, \cdots, 2N$

The coefficient d_m for constructing $E^e(t)$ can be obtained from

$$[d_m] = \begin{bmatrix} M_{nm}^c \\ \cdots \\ M_{nm}^s \end{bmatrix}^{-1} \begin{bmatrix} A_n \\ \vdots \\ B_n \end{bmatrix} \tag{5.10}$$

To synthesize $E^e(t)$ for the jth-mode extraction, we assign

$$A_j = 1 \text{ and } B_j = 1$$

and all other A_n and $B_n = 0$

$$[d_m] = \begin{bmatrix} M_{nm}^c \\ \vdots \\ M_{nm}^s \end{bmatrix}^{-1} \begin{bmatrix} 0 \\ \vdots \\ 1 \\ \vdots \\ 0 \\ \vdots \\ 1 \\ \vdots \\ 0 \end{bmatrix} \begin{matrix} \\ \\ \leftarrow A_j \\ \\ \\ \\ \leftarrow B_j \\ \\ \\ \end{matrix} \tag{5.11}$$

Then select an appropriate short duration for T_e in such a way that the synthesized $E^e(t)$ is a well behaved waveform.

To synthesize an zero-mode extraction signal so that the convolved radar output is zero, we let

$$A_n = B_n = 0 \qquad \text{for all } n$$

In this case $[d_m]$ will have nontrivial solutions only when the determinant of $\begin{bmatrix} M_{nm}^c \\ \vdots \\ M_{nm}^s \end{bmatrix}$ vanishes. That is

$$det \begin{bmatrix} M_{nm}^c \\ \vdots \\ M_{nm}^s \end{bmatrix} = 0 \tag{5.12}$$

This condition can be met because all the elements of M_{nm}^c and M_{nm}^s are functions of the signal duration T_e, and it is possible to numerically search for the optimum value of T_e to satisfy the stated condition. Once the optimum T_e is determined, $[d_m]$ can be easily determined from a set of homogeneous equations generated from Eq. (5.9).

The discriminant signals for zero response or single-mode response were synthesized based on the prior knowledge of target's natural frequencies. In the case of complex targets, this information is difficult to obtain, and the synthesis of discriminant signals will be based on an experimental measurement of the pulse response or the late-time response of the scale model of the target combined with some theoretical techniques such as continuation method [7] or the fast Prony's method [8].

EXPERIMENTAL VERIFICATION OF E-PULSE METHOD*

The feasibility and accuracy of the E-pulse scheme were verified experimentally. A series of experiments were conducted at the MSU transient EM scattering facility, which includes a ground plane scattering range and anechoic chamber scattering range with associated electronics components, as schematically depicted in Fig. 1.

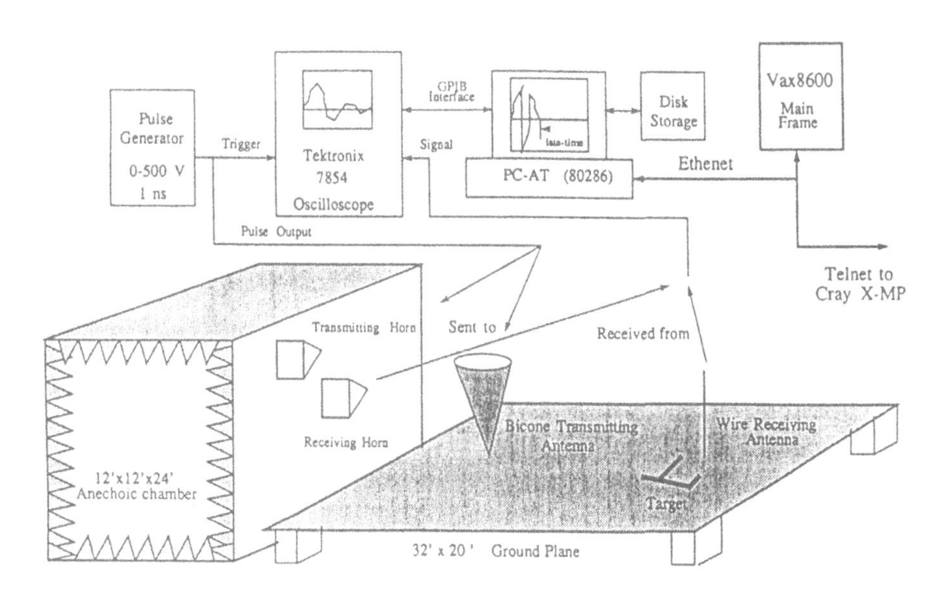

Fig. 1 Two scattering ranges with associated equipment in the Electromagnetics Laboratory of Michigan State University.

*The material in this section is based on "New progress on E/S pulse techniques for noncooperative target recognition" by Kun-Mu Chen, C. P. Nyquist, E. Rothwell and W. M. Sun, which appeared in IEEE Transactions on Antennas and Propagation, Vol. 40, No. 7, pp. 829-833, July 1992. (©1992 IEEE.)

The ground plane time-domain scattering range implements antennas and target models imaged on a 32'×20' ground plane. A large monoconical antenna, with height 2.4 m, apex angle 16°, and characteristic impedance 160 Ω, is used as the transmitting antenna to radiate nanosecond EM pulses. A long wire antenna of length 2.4 m and diameter 0.635 cm is used as the receiving antenna to receive the scattered fields from the targets. Scale models of various airplanes have been constructed and they can be placed at various locations on the ground plane to be illuminated by the EM pulse radiated by the transmitting antenna, and their scattered fields received by the receiving antenna. A nanosecond EM pulse generator (Tektronix 109 mercury-switched pulse generator) is used to excite the transmitting antenna. This pulse generator produces pulses of 100 *ps* risetime and duration between 1 *ns* and 1 *μs* with amplitudes as great as 500 V at 1 KHz repetition rate.

The scattered field or pulse response from targets are sampled and measured using a Tektronix 7854 digital waveform processing sampling oscilloscope (DWPSO). The DWPSO automatically acquires and averages multiple target response waveforms, and implement initial signal processing operation such as interpolation, smoothing and integration. An IEEE GPIB interface links the DWPSO with an IBM-AT compatible microcomputer. Software programs executed on the microcomputer transfer data from the DWPSO waveform memories to compute RAM, and subsequently to hard-disk storage. After the measured scattered fields from the targets are transferred from the DWPSO to the computer, they are numerically convolved with the discriminant signals of the targets stored in the computer. The convolved outputs are then displayed on the computer monitor.

We aimed to discriminate and then identify among four different target models: 1) a medium size B707 model of length 33 cm, 2) a medium size T-15 model (homemade and arbitrarily named) of length 30 cm, 3) a big B707 model of length 64.5 cm, and 4) a big F-18 model of length 72 cm as shown in Fig. 2. The E-pulses of these four models have been synthesized as shown in Fig. 3, and they were stored in the computer. The scattered fields or the pulses responses of these four target measured at certain aspect angles are shown in Fig. 4. It is observed that these scattered fields consist of large early-time responses followed by oscillatory late-time response. The shapes of these scattered fields are strongly dependent on the aspect angle and it is impossible to identify the targets from the scattered fields. However, when these scattered fields are convolved with the E-pulses of four different targets in such a way as depicted as Fig. 2, the targets can be easily and clearly identified.

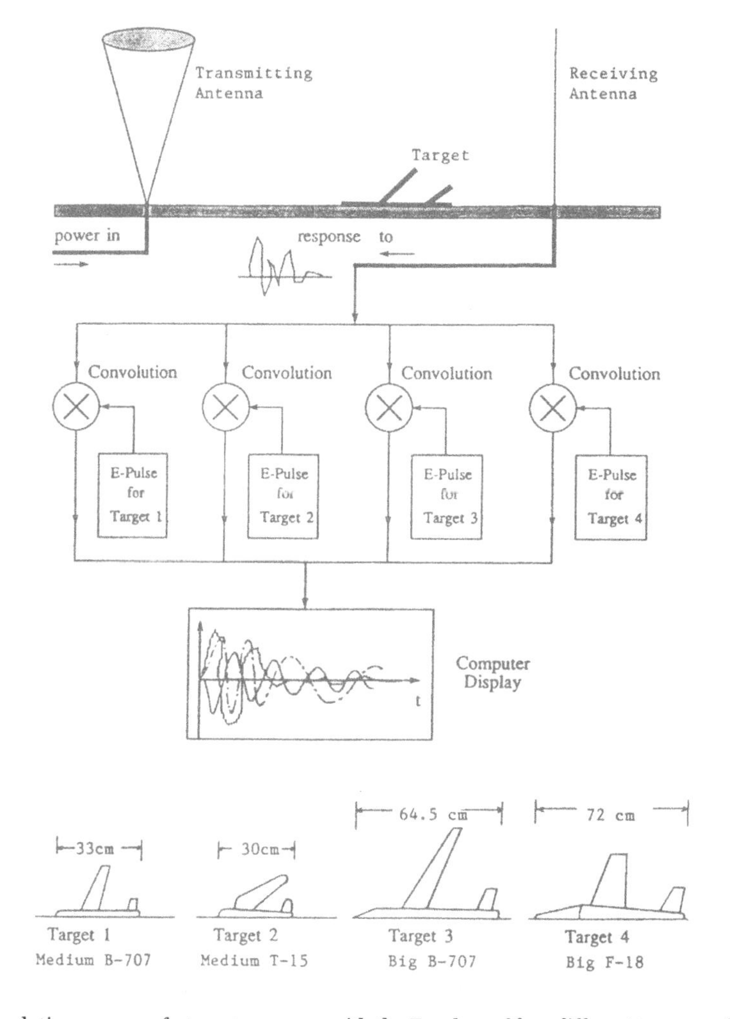

Fig. 2 Convolution process of a target response with the E-pulses of four different target models.

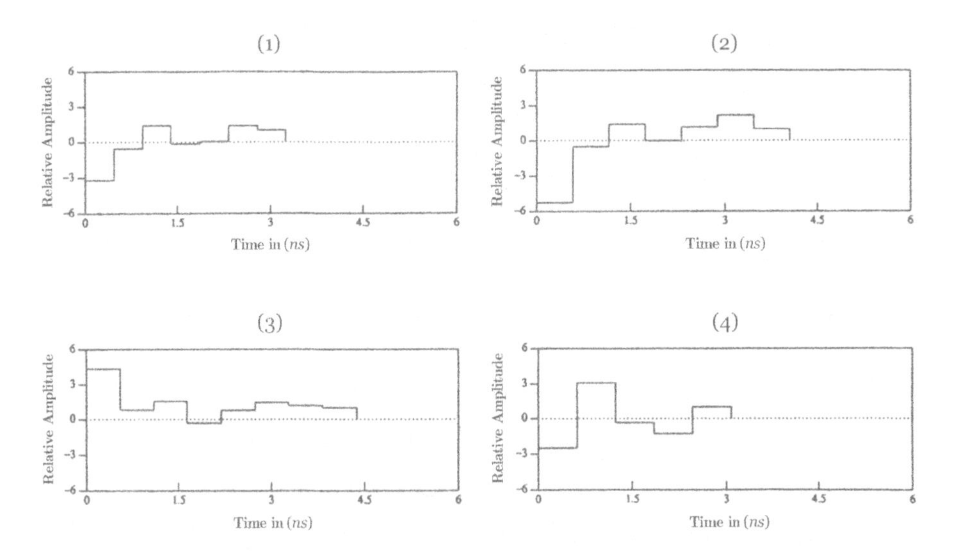

Fig. 3 E-pulses for the four target models: (1) a medium size B-707 model, (2) a medium size T-15 model, (3) a big B-707 model, and (4) a big F-18 model.

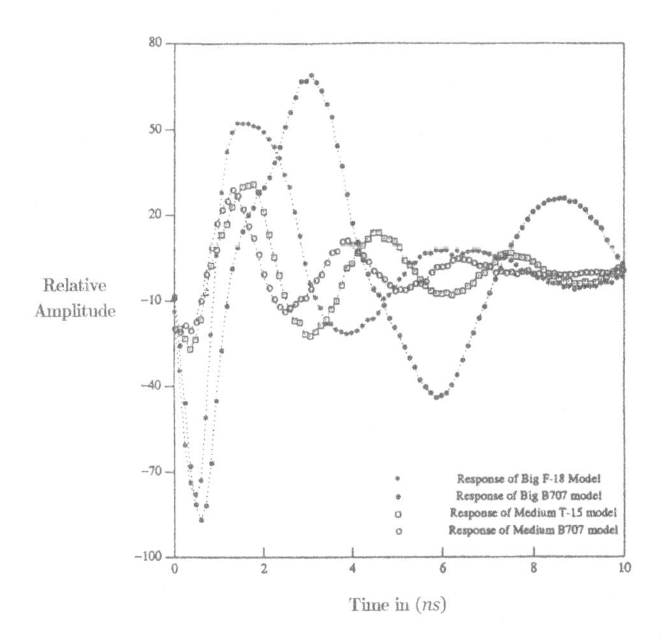

Fig. 4 The scattered fields or the pulse responses of the four target models measured at certain aspect angles.

Fig. 5 shows the four convolved out put signals when the scattered field of the medium B707 model was convolved with four E-pulses of the four different targets. It is observed that the convolved output of this scattered field with E-pulse of the medium B707 model (the same target) gives a very small signal (almost a flat line) in the late-time period. On the other hand, three other targets (wrong targets) all give large late-time responses. Thus, it is very easy to identify from these results that the scattered field belongs to the medium B-707 model.

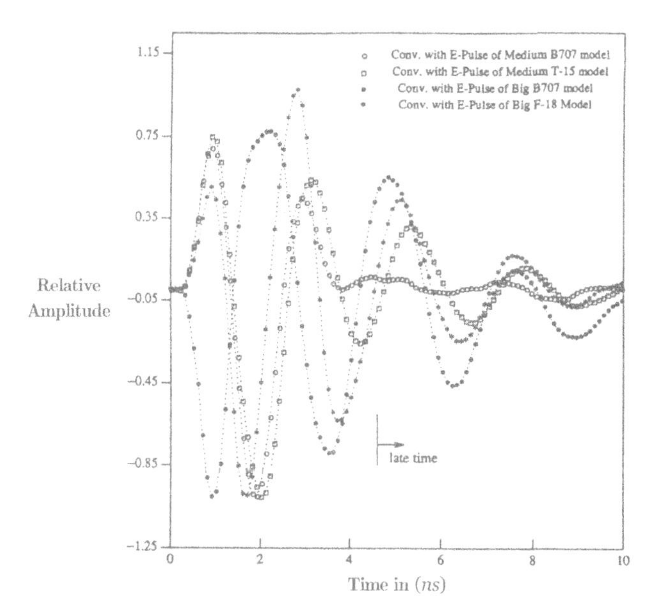

Fig. 5 Convolution of the scattered field of the medium B-707 model with the E-pulses of medium B-707 model, medium T-15, model, big B-707 model and big F-18 model.

Fig. 6 shows the four convolved output signals when the scattered field of the medium T-15 models was convolved with four E-pulses of the four different targets. From the convolved output signal with a very small late-time response (almost a flat line), which is the convolved output of the scattered field with the E-pulse of the medium T-15 model, it is easy to identify that the measured target is the medium T-15 models.

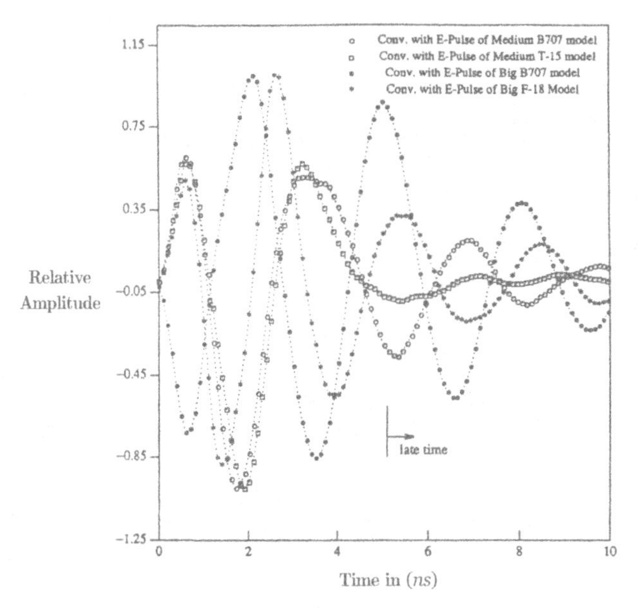

Fig. 6 Convolution of the scattered field of the medium T-15 model with the E-pulses of medium B-707 model, medium T-15 model, big B-707 model and big F-18 model.

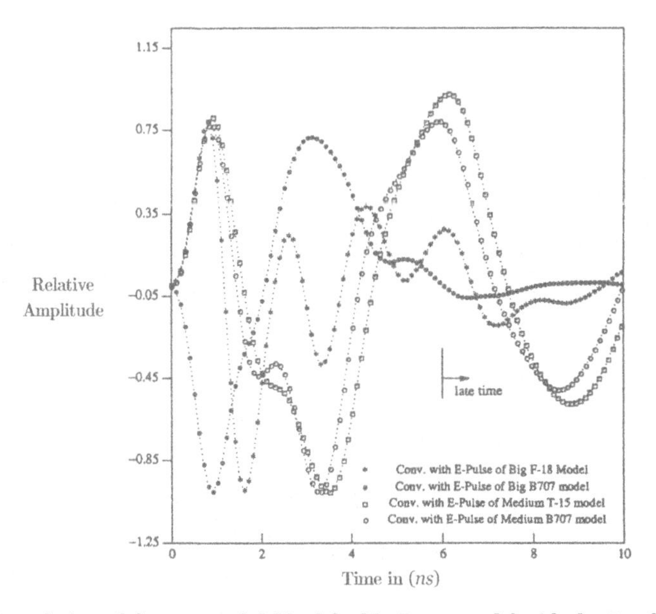

Fig. 7 Convolution of the scattered field of the big B-707 model with the E-pulses of medium B-707 model, medium T-15 model, big B-707 model and big F-18 model.

Fig. 7 shows the four convolved output signals when the scattered field of the big B-707 model was convolved with the E-pulses of the four different targets. Again it is very easy to identify the target being measured as the big B-707 model from the convolved output of this scattered field with the E-pulse of the big B707 model. Fig. 8 shows similar results when the scattered field of the big F-18 was convolved with the E-pulses of the four different targets. From the convolved output signal with a flat late-time response, the target in question can be easily identified as the big F-18 model.

We have tested the aspect-independency of the scheme by varying the target's aspect angle. Similar results of Figs. 5-8 were obtained for each case of aspect angle.

In the course of the study of the E-pulse method, it was found that the E-pulse is also noise-insensitive, in addition to being aspect-independent.

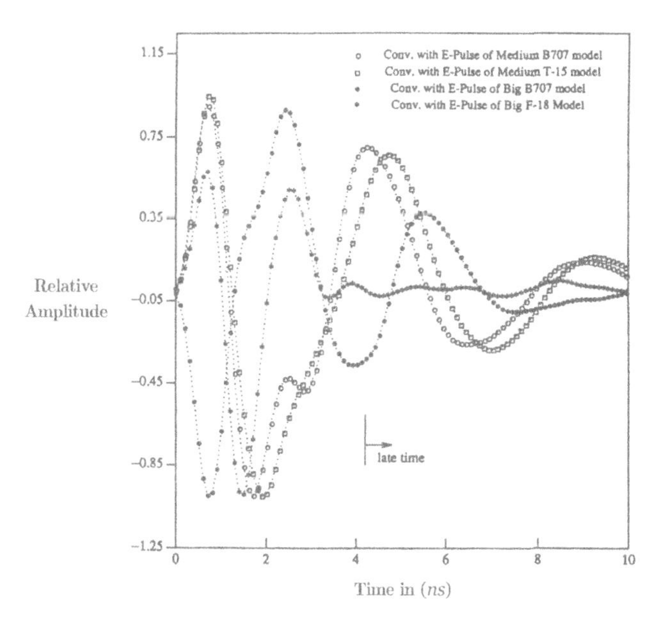

Fig. 8 Convolution of the scattered field of the big F-18 model with the E-pulses of medium B-707 model, medium T-15 model, big B-707 model and big F-18 model.

To prove the noise insensitivity of the E-pulse scheme, we have created very noisy radar responses of complex targets by intentionally adding a large random noise to the measured radar response of the targets. These noisy responses were then used to convolve with the E-pulses of the targets. We have found that the E-pulse signals of targets are very powerful and effective in rejecting a large random noise and are capable of discriminating between the right and the wrong targets from very noisy radar responses. The following figures will demonstrate this finding.

Fig. 9 shows the pulse response of the B707 model measured at 90° aspect angle without an extra random noise added. Fig. 10 is the convolved output of the pulse response of Fig. 9 with the E-pulse of the B707 model. As expected, a very small output was obtained in the late-time period of the convolved output. This indicated that the pulse response of Fig. 9 came from the right target of the B707 model. Next, a very noisy pulse response was created by intentionally adding a large random noise (created by a computer) to the measured pulse response of the B707 model shown in Fig. 9. This random noise amounted to 30% of the maximum amplitude of the measured early-time response of Fig. 9. It is noted that the added random noise has a flat wide frequency spectrum and it cannot be simply filtered out by a low-pass filter. The created noisy pulse response is shown in Fig. 11. When this noisy pulse response of Fig. 11 was convolved with the E-pulse of the B707 models, a very satisfactory convolved output was obtained, as shown in Fig. 12. This convolved output resembles that of Fig. 10; the early-time response stayed nearly unchanged, and more importantly, the late-time response still remained small. This indicated that the E-pulse of the B707 model was capable of identifying the noisy pulse response of Fig. 11 belonging to the B707 model.

These results indicate the noise insensitivity of the E-pulse technique. The reason for this attractive characteristic can be attributed to the convolution process, a signal averaging process in the E-pulse technique.

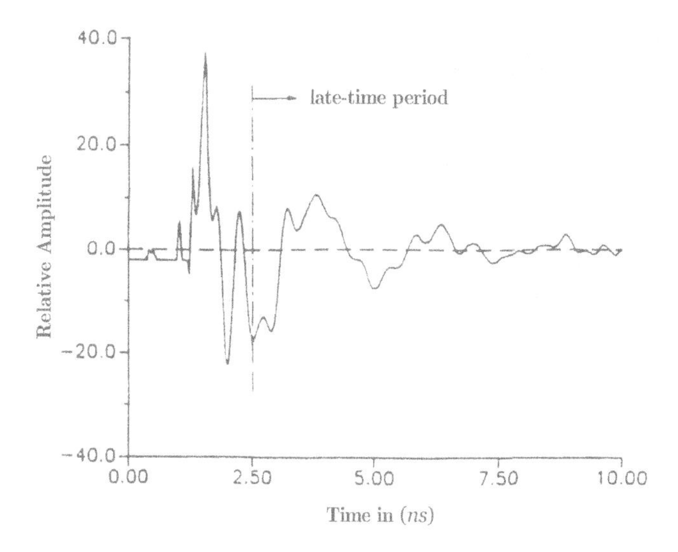

Fig. 9 Pulse response of B-707 model measured at 90° aspect angle without an extra noise added.

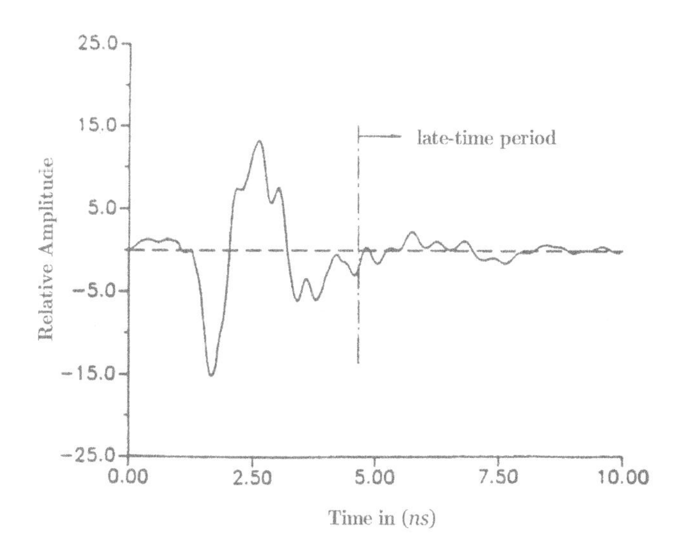

Fig. 10 Convolved output of the E-pulse of B-707 model with the pulse response of B-707 model measured at 90° aspect without an extra random noise added. A very small late-time response was obtained.

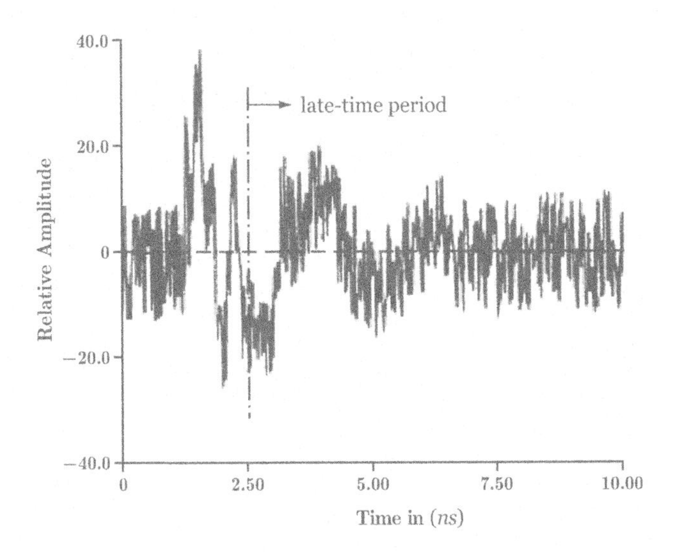

Fig. 11 Pulse response of B-707 model measured at 90° aspect angle with an extra random noise (30% of the maximum waveform amplitude) added.

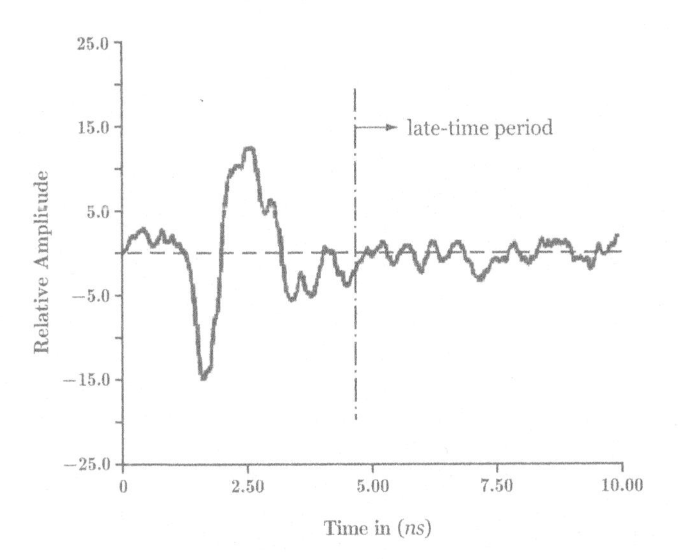

Fig. 12 Convolved output of the E-pulse of B-707 model with the pulse response of B-707 model measured at 90° aspect angle with an extra random noise (30% maximum amplitude) added. The late-time response was still very small.

 E-Pulse Method — Frequency Domain Analysis

In the preceding section, the E-pulse method, an aspect-independent radar target identification scheme, was analyzed in the time-domain approach. In this section, the E-pulse method will be developed with a frequency-domain analysis.

The target response during the late-time period can be written as

$$r(t)=\sum_{n=1}^{N}a_n e^{\sigma_n t}\cos(\omega_n t+\phi_n) \qquad \text{for } t>T_e \tag{5.13}$$

The convolution of an E-pulse, $e(t)$, with the late-time target response gives

$$\begin{aligned}
c(t) &= e(t)*r(t) && \text{— Convolution of } e(t) \text{ and } r(t)\\
&= \int_0^t e(t')r(t-t')dt' && (5.14)\\
&= \int_0^{T_e} e(t')r(t-t')dt' && \text{because } e(t')=0 \text{ for } t'>T_e \text{ and } t'<0
\end{aligned}$$

Taking the Laplace transform of $c(t)$:

$$L[c(t)]=L[e(t)*r(t)]=L\left[\int_0^t e(t')r(t-t')dt'\right]=E(s)R(s) \tag{5.15}$$

where

$$E(s)=L[e(t)]=\int_0^\infty e(t)e^{-st}dt \tag{5.16}$$

$$\begin{aligned}
R(s) &= L[r(t)]=\int_0^\infty r(t)e^{-st}dt\\
&= \int_0^\infty \sum_{n=1}^{N} a_n e^{\sigma_n t}\cos(\omega_n t+\phi_n)e^{-st}dt\\
&= \int_0^\infty \sum_{n=1}^{N}\frac{1}{2}a_n e^{\sigma_n t}\left[e^{i(\omega_n t+\phi_n)}+e^{-i(\omega_n t+\phi_n)}\right]e^{-st}dt
\end{aligned}$$

$$
\begin{aligned}
R(s) &= \int_0^\infty \sum_{n=1}^{N} \frac{1}{2} a_n \left[e^{(\sigma_n + i\omega_n)t} e^{i\phi_n} + e^{(\sigma_n - i\omega_n)t} e^{-i\phi_n} \right] e^{-st} dt \\
&= \int_0^\infty \sum_{n=1}^{N} \frac{1}{2} a_n \left[e^{s_n t} e^{i\phi_n} + e^{s_n^* t} e^{-i\phi_n} \right] e^{-st} dt \\
&= \int_0^\infty \sum_{n=1}^{N} \frac{1}{2} a_n \left[e^{i\phi_n} e^{-(s-s_n)t} + e^{-i\phi_n} e^{-(s-s_n^*)t} \right] dt \\
&= \sum_{n=1}^{N} -\frac{1}{2} a_n \left[e^{i\phi_n} \frac{e^{-(s-s_n)t}}{s-s_n} \Bigg|_0^\infty + e^{-i\phi_n} \frac{e^{-(s-s_n^*)t}}{s-s_n^*} \Bigg|_0^\infty \right] \\
&= \sum_{n=1}^{N} \frac{1}{2} a_n \left[\frac{e^{i\phi_n}}{s-s_n} + \frac{e^{-i\phi_n}}{s-s_n^*} \right]
\end{aligned} \tag{5.17}
$$

$$
E(s) = L[e(t)] = \int_0^\infty e(t) e^{-st} dt
$$

$$
\begin{aligned}
E(s_n) &= \int_0^\infty e(t) e^{-s_n t} dt = \int_0^\infty e(t) e^{-(\sigma_n + i\omega_n)t} dt \\
&= \int_0^\infty e(t) e^{-\sigma_n t} e^{-i\omega_n t} dt \\
&= \int_0^\infty e(t) e^{-\sigma_n t} \cos(\omega_n t) dt - i \int_0^\infty e(t) e^{-\sigma_n t} \sin(\omega_n t) dt \\
&= \mathrm{Re}[E(s_n)] - i\,\mathrm{Im}[E(s_n)] = E_{rn} - iE_{in} \\
&= |E(s_n)| e^{-i\tan^{-1}(E_{in}/E_{rn})} = |E(s_n)| e^{i\tan^{-1}(-E_{in}/E_{rn})} \\
&= |E(s_n)| e^{i\theta_n} \qquad \text{where } \theta_n = \tan^{-1}(-E_{in}/E_{rn}) \\
E(s_n^*) &= |E(s_n)| e^{-i\theta_n}
\end{aligned} \tag{5.18}
$$

Since $\quad L[c(t)] = E(s)R(s)$

$$
\begin{aligned}
c(t) &= L^{-1}[E(s)R(s)] \\
&= \frac{1}{2\pi i} \int_{r-i\infty}^{r+i\infty} E(s)R(s) e^{st} ds \\
&= \frac{1}{2\pi i} \int_{r-i\infty}^{r+i\infty} \sum_{n=1}^{N} \frac{1}{2} a_n \left[\frac{e^{i\phi_n} E(s) e^{st}}{s-s_n} + \frac{e^{-i\phi_n} E(s) e^{st}}{s-s_n^*} \right] ds \\
&= \sum_{n=1}^{N} \frac{1}{2} a_n \left[e^{i\phi_n} E(s_n) e^{s_n t} + e^{-i\phi_n} E(s_n^*) e^{s_n^* t} \right] \\
&= \sum_{n=1}^{N} \frac{1}{2} a_n \left[e^{i\phi_n} |E(s_n)| e^{i\theta_n} e^{s_n t} + e^{-i\phi_n} |E(s_n)| e^{-i\theta_n} e^{s_n^* t} \right] \\
&= \sum_{n=1}^{N} \frac{1}{2} a_n \left[|E(s_n)| e^{\sigma_n t} e^{i(\omega_n t + \phi_n + \theta_n)} + |E(s_n)| e^{\sigma_n t} e^{-i(\omega_n t + \phi_n + \theta_n)} \right] \\
&= \sum_{n=1}^{N} a_n |E(s_n)| e^{\sigma_n t} \cos(\omega_n t + \phi_n + \theta_n) \qquad t > T_e
\end{aligned} \tag{5.19}
$$

where $|E(s_n)| = \left[E_{rn}^2 + E_{in}^2\right]^{1/2}$ and $\theta_n = \tan^{-1}(-E_{in} / E_{rn})$

The convolved output of the E-pulse with the late-time radar target response given in Eq. (5.19) looks similar to the corresponding result in the time-domain analysis given in Eq. (5.15) of the preceding section, but they are different. Therefore, the synthesis of the E-pulse signal in the frequency-domain will be somewhat different from that given in the preceding section, as given below.

We can proceed to synthesize the E-pulse signals.

Now, to have zero convolved radar return,

$$c(t)=0 \qquad \text{for } t > T_e$$

it requires that

$$E_{rn} = E_{in} = 0 \qquad \text{for } 1 \le n \le N \tag{5.20}$$

Or equivalently,

$$E(s_n) = E(s_n^*) = 0 \qquad \text{for } 1 \le n \le N \tag{5.21}$$

For a single-mode (m^{th} mode) extraction,

$$E(s_n) = E(s_n^*) = 0 \qquad \text{for } 1 \le n \le N\,,\ n \ne m \tag{5.22}$$

While for a sin m^{th} mode extraction,

$$\text{and} \begin{cases} E(s_n) = E(s_n^*) = 0 & \text{for } 1 \le n \le N,\ n \ne m \\ E(s_m) = -E(s_m^*) & \end{cases} \tag{5.23}$$

For a cosine m^{th} mode extraction,

$$\text{and} \begin{cases} E(s_n) = E(s_n^*) = 0 & \text{for } 1 \le n \le N,\ n \ne m \\ E(s_m) = E(s_m^*) & \end{cases} \tag{5.24}$$

To implement the E-pulse, we can also add a forcing component to the E-pulse such as

$$e(t) = e^f(t) + e^e(t) \tag{5.25}$$

where $e^f(t)$ is a forcing component which excites the target, and $e^e(t)$ is an extinction component which extinguishes the response due to $e^f(t)$. The forcing component is chosen freely, while the extinction component is determined by first expanding into a set of basis functions,

$$e^e(t) = \sum_{m=1}^{M} \alpha_m f_m(t) \tag{5.26}$$

$$\begin{aligned} E(s_n) &= \int_0^\infty e(t)e^{-s_n t}dt = \int_0^\infty \left[e^f(t) + e^e(t)\right]e^{-s_n t}dt \\ &= \int_0^\infty e^f(t)e^{-s_n t}dt + \int_0^\infty e^e(t)e^{-s_n t}dt \\ &= E^f(s_n) + \sum_{m=1}^{M} \alpha_m \int_0^\infty f_m(t)e^{-s_n t}dt \\ &= E^f(s_n) + \sum_{m=1}^{M} \alpha_m F_m(s_n) \end{aligned} \tag{5.27}$$

To have

$$E(s_n) = E(s_n^*) = 0 \qquad \text{for } 1 \le n \le N$$

$$\begin{bmatrix} F_1(s_1) & F_2(s_1) & \cdots & F_M(s_1) \\ F_1(s_2) & F_2(s_2) & \cdots & F_M(s_2) \\ & \cdots & & \\ F_1(s_1^*) & F_2(s_1^*) & \cdots & F_M(s_1^*) \\ & \cdots & & \\ F_1(s_N^*) & F_2(s_N^*) & \cdots & F_M(s_N^*) \end{bmatrix} \begin{bmatrix} \alpha_1 \\ \alpha_2 \\ \vdots \\ \alpha_M \end{bmatrix} = \begin{bmatrix} -E^f(s_1) \\ -E^f(s_2) \\ \vdots \\ -E^f(s_1^*) \\ \vdots \\ -E^f(s_N^*) \end{bmatrix} \tag{5.28}$$

where $F_m(s) = L\left[f_m(t)\right]$

$E^f(s) = L\left[e^f(t)\right]$ and $M=2N$.

If we use the pulse basis functions :

$$e^{e}(t)=\sum_{m=1}^{M}\alpha_{m}f_{m}(t)$$

$$f_{m}(t)=\begin{cases} g(t-(m-1)\Delta) & (m-1)\Delta\leq t\leq m\Delta \\ 0 & elsewhere \end{cases}$$

$$F_{m}(s)=\int_{0}^{\infty}g(t-(m-1)\Delta)e^{-st}dt$$

$$=F_{1}(s)e^{s\Delta}e^{-sm\Delta} \qquad (5.29)$$

where

$$F_{1}(s)=\int_{0}^{\infty}g(t)e^{-st}dt$$

EXPERIMENTAL RESULTS

Some experimental results are given here to show the discrimination between two similar aircraft models using single-mode extraction signals.

Two experimental targets are F-18 and B707 aircraft models which were described in the preceding section. The dominant frequencies of these models were extracted from the late-time portion of the responses using a Continuation method [7]. With these, pulse function based natural sine and cosine single-mode extraction waveforms were constructed.

For example, Fig. 13 shows the first mode extraction waveforms for B707 model.

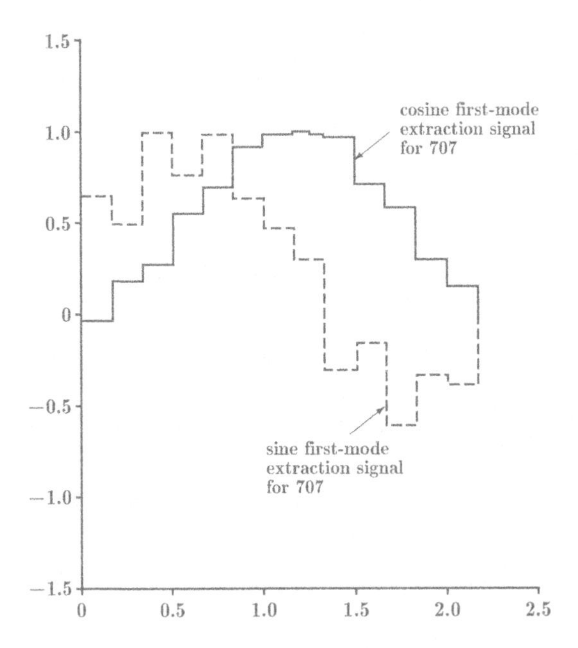

Fig. 13 Natural rectangular pulse function based first mode extraction waveforms for the Boeing 707 aircraft model.

Reprinted from "Frequency domain E-pulse synthesis and target discrimination," by E. J. Rothwell, K. M. Chen, D. P. Nyquist and W. Sun, IEEE Trans. on Antennas and Propagation, Vol. AP-35, No. 4, pp. 426-434, April 1987. (©1987 IEEE.)

When the late-time target response is convolved with the sine m^{th} mode extraction signal, it will yield an output as

$$C(t) = C_S(t) = a_m e^{\sigma_m t} B_m \sin(\omega_m t + \phi_m) \tag{5.30}$$

And the convolution of the late-time target response with the cosine m^{th} mode extraction signal will give an output of

$$C(t) = C_C(t) = a_m e^{\sigma_m t} A_m \cos(\omega_m t + \phi_m) \tag{5.31}$$

With the proper normalization of the extraction signals, we can make $A_m = B_m$. Then Eqs. (5.30) and (5.31) can be combined to yield plots of the mth mode frequencies versus time, via

$$\begin{aligned} \omega_m t + \phi_m &= \tan^{-1}\frac{C_S(t)}{C_C(t)} \\ \sigma_m t + \log|a_m A_m| &= \frac{1}{2}\log\left[C_C^2(t) + C_S^2(t)\right] \end{aligned} \tag{5.32}$$

In this example, the first and fourth mode extraction signals for the B707 model ware synthesized. These single-mode extraction signals were then convolved with the late-time target response of B707 and F-18 models. Convolved outputs were then plotted as functions of time using Eq. (5.32). If the single-mode extraction signals for B707 model are convolved with the expected B707 response, the result should be either a pure first or fourth mode damped sinusoid. If these extraction signals of B707 model are convolved with unexpected F-18 model response, the result will be unrecognizable conglomeration of the modes of the unexpected target.

Figs. 14-17 show the results of convolving the B707 extraction signals with both the B707 and F-18 model response. These frequency plots are obtained from the actual convolved waveforms using Eq. (5.32). The dotted lines represent the slopes of the expected first or fourth mode frequencies. It is seen that the frequency plots for the expected B707 target (Figs. 14 and 15) parallel the expected frequency lines in the late-time period, while those for the unexpected F-18 target (Figs. 16 and 17) do not. Thus, the B707 target and the F-18 targets are easily discriminated.

A POTENTIAL RADAR IDENTIFICATION SYSTEM BASED ON THE E-PULSE METHOD

Fig. 18 depicts a potential radar detection system based on the present scheme of convolving the target radar return with the synthesized discriminant signals. The system consists of a network of computers and each of them is assigned to store the synthesized signal for extracting various single-mode or zero-mode response for a particular friendly target. All the relevant friendly

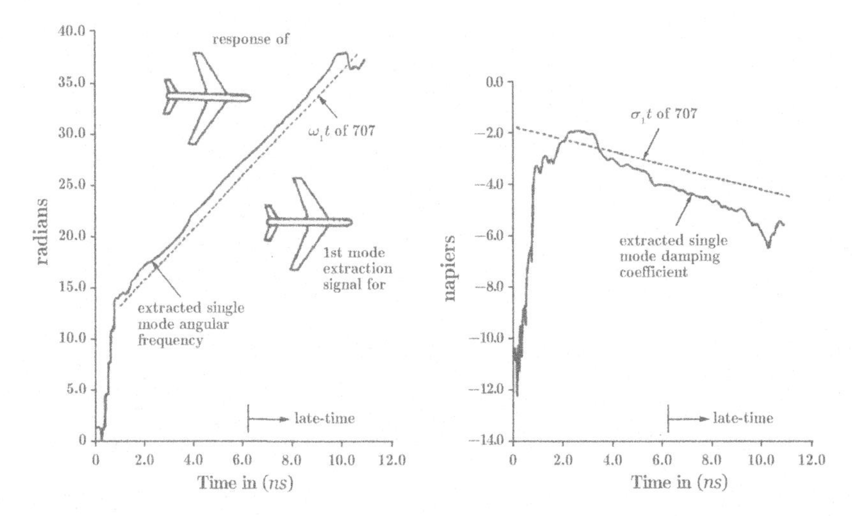

Fig. 14 **Single mode angular frequency and damping coefficient extracted from the convolved outputs of the first mode extraction signals for the 707 target and the 707 measured response.**

Reprinted from "Frequency domain E-pulse synthesis and target discrimination," by E. J. Rothwell, K. M. Chen, D. P. Nyquist and W. Sun, IEEE Trans. on Antennas and Propagation, Vol. AP-35, No. 4, pp. 426-434, April 1987. (©1987 IEEE.)

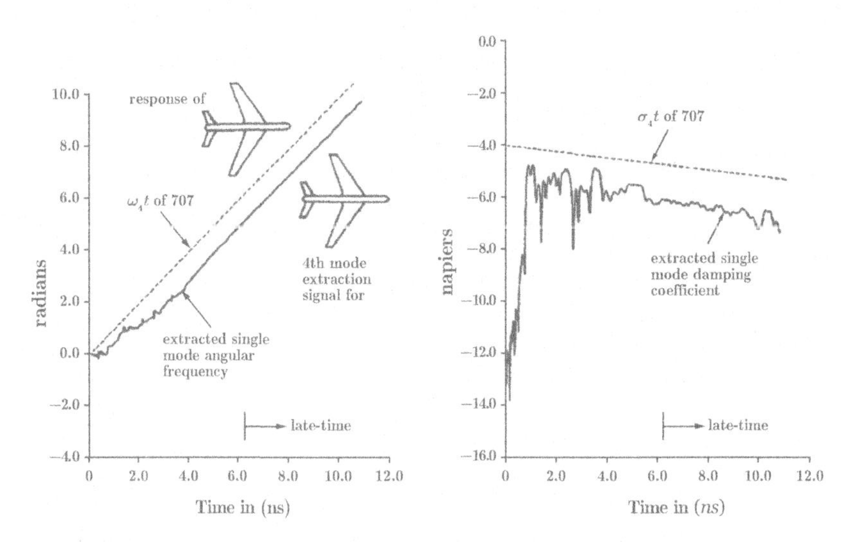

Fig. 15 **Singular mode angular frequency and damping coefficient extracted from the convolved outputs of the fourth mode extraction signals for the 707 target and the 707 measured response.**

Reprinted from "Frequency domain E-pulse synthesis and target discrimination," by E. J. Rothwell, K. M. Chen, D. P. Nyquist and W. Sun, IEEE Trans. on Antennas and Propagation, Vol. AP-35, No. 4, pp. 426-434, April 1987. (©1987 IEEE.)

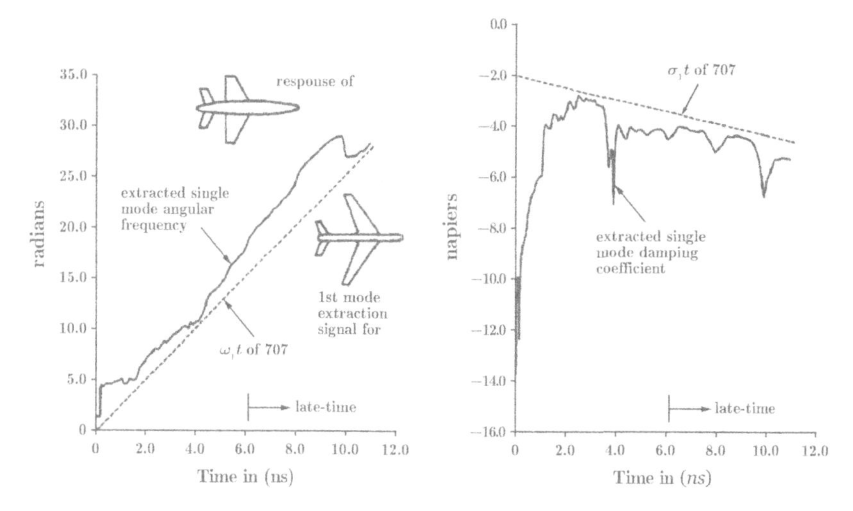

Fig. 16 Single mode angular frequency and damping coefficient extracted from the convolved outputs of the first mode extraction signals for the 707 target and the F-18 measured response.

Reprinted from "Frequency domain E-pulse synthesis and target discrimination," by E. J. Rothwell, K. M. Chen, D. P. Nyquist and W. Sun, IEEE Trans. on Antennas and Propagation, Vol. AP-35, No. 4, pp. 426-434, April 1987. (©1987 IEEE.)

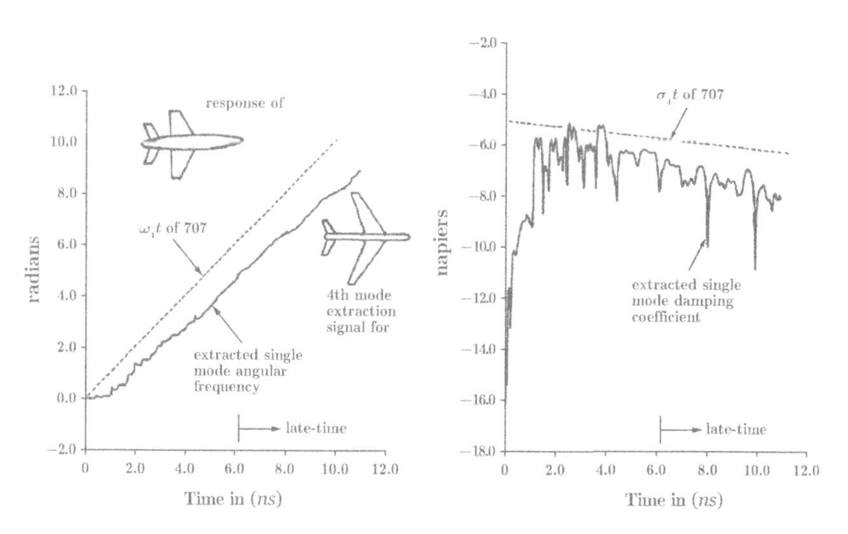

Fig. 17 Single mode angular frequency and damping coefficient extracted from the convolved outputs of the fourth mode extraction signals for the 707 target and the F-18 measured response.

Reprinted from "Frequency domain E-pulse synthesis and target discrimination," by E. J. Rothwell, K. M. Chen, D. P. Nyquist and W. Sun, IEEE Trans. on Antennas and Propagation, Vol. AP-35, No. 4, pp. 426-434, April 1987. (©1987 IEEE.)

targets are assumed to be covered in the network of computers. When an approaching target is illuminated by an interrogating radar signal, the radar return is divided and fed to each computer after amplification and signal processing. Inside each computer the stored discriminant signals are convolved with the radar return. In principle, only one of the computers will produce various single-mode and zero-mode outputs in the late-time period; the rest of the computers should produce irregular outputs. The computer producing the single-mode and zero-mode outputs will then be identified with the target. If none of the computers produces single-mode and zero outputs in the late-time period, the approaching targets will not be friendly.

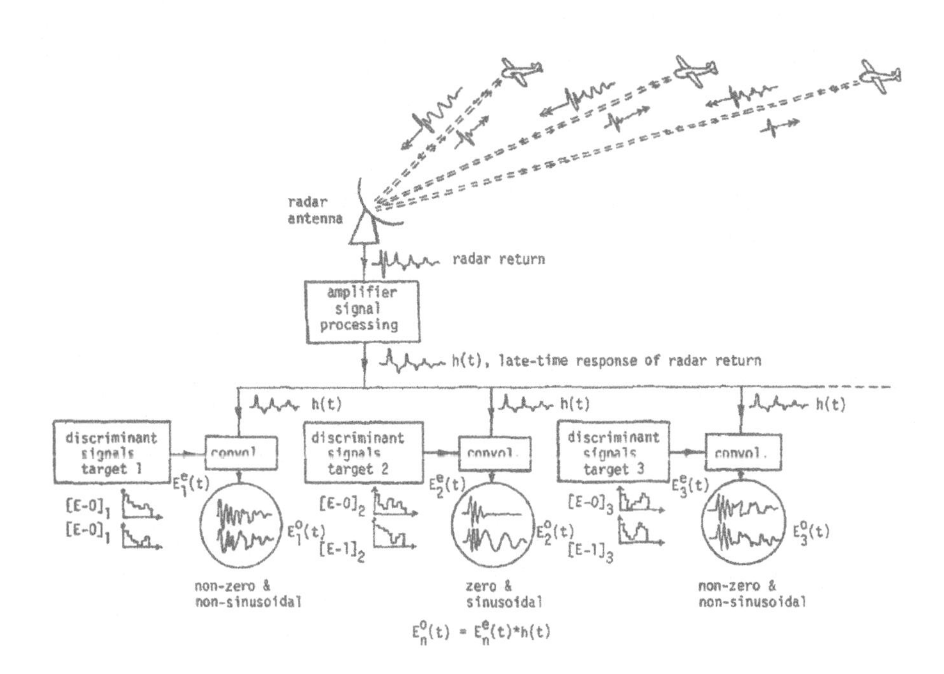

Fig. 18 A proposed target discrimination and identification system.

 Application of E-pulse Technique to Early-Time Target Response

Preceding two sections describe the aspect-independent target identification method using the late-time radar return and the E-pulse technique. This section will study the possibility of applying the E-pulse technique to the early-time radar return for the target identification.

The late-time radar return contains the information of target's natural frequencies which are aspect-independent. That is why the application of the E-pulse technique to the late-time radar return can provide the aspect-independent target identification. The early-time radar return contains mainly specular reflections from target scattering centers which are aspect-dependent. However, the early-time part of the radar return contains more energy than the late-time part. Therefore, it is desirable if the early-time radar return can be utilized for the target identification. Fortunately, the E-pulse technique is usable with waveforms arising from specular scattering, such as the early-time response of a radar target. The technique uses resonance cancellation in the frequency domain to eliminate the sinusoidal functions arising from the aspect-dependent temporal positions of specular reflections. Target discrimination using early-time information is then possible using an algorithm identical to that used with late-time data, with the exception that discrimination is aspect-dependent.

SYNTHESIS OF E-PULSE FOR TARGET IDENTIFICATION USING EARLY-TIME RESPONSE

The possible E-pulse method for the target identification using the early-time portion of the target response is analyzed here.

Figure 19 shows the transient response of a 1:72 scale model of a B-52 aircraft to an excitation pulse with energy in the band 0.2-7.0 GHz. It is typical of the return from an aircraft target, showing an early-time period (3.0-7.0 *ns*) dominated by localized specular reflections from scattering centers such as engine intakes and attachment points, followed by a late-time natural oscillation period consisting of global resonance information ($t > 7.0$ *ns*). It is important to note that there can be no precise demarcation between the early-time and late-time portions of a scattered field response since substructure resonances are often established before the excitation signal clear the target, resulting in a resonance component to early time.

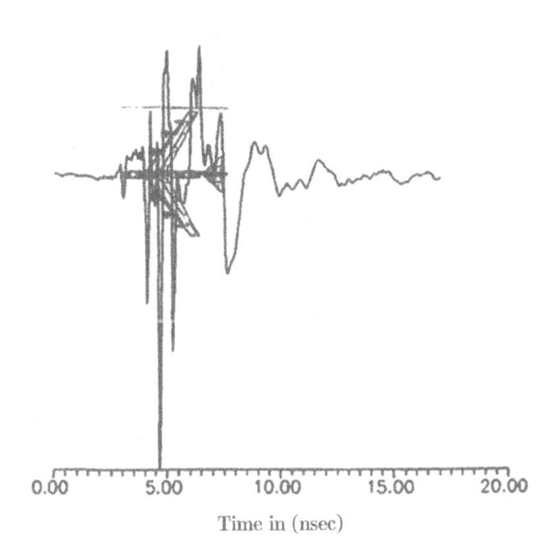

Fig. 19 Response of 1:72 scale B-52 model measured at noise-on incidence in the frequency band 0.2-7.0 GHz. Polarization is in plane of wings.

The late-time target response can be expressed as

$$\begin{aligned} r_L(t) &= \sum_{n=1}^{N} a_n e^{\sigma_n t} \cos(\omega_n t + \phi_n) \\ &= \sum_{n=-N}^{N} A_n e^{S_n t} = \sum_{n=-N}^{N} a_n e^{i\phi_n} e^{(\sigma_n + i\omega_n)t} \\ &\text{for } t > T_L \end{aligned} \tag{5.33}$$

where

$A_n = a_n e^{i\phi_n}$, $A_{-n} = a_n e^{-i\phi_n}$

$S_n = \sigma_n + i\omega_n$, $S_{-n} = \sigma_n - i\omega_n$

$T_L =$ beginning of late-time.

$N =$ number of natural frequencies in conjugate pairs.

The early-time target response can be expressed as

$$r_E(t) = \sum_{m=1}^{M} p(t) * h_m(t - T_m) \qquad t < T_L \tag{5.34}$$

where

$p(t) =$ incident pulse

$h_m(t) =$ localized impulse response originating at the mth scattering center at the time T_m.

In the frequency domain $r_E(t)$ becomes, after taking the Fourier transform:

$$\begin{aligned} R_E(\omega) &= F\left[r_{E(t)}\right] = \frac{1}{\sqrt{2\pi}} \int_{-\infty}^{\infty} r_E(t) e^{-i\omega t} dt \\ &= \sum_{m=1}^{M} P(\omega) H_m(\omega) e^{-i\omega T_m} \end{aligned} \tag{5.35}$$

where

$P(\omega) = spectrum$ of $P(t) = \dfrac{1}{2\pi} \displaystyle\int_{-\infty}^{\infty} P(t) e^{-i\omega t} dt$

$H_m(\omega) =$ transfer function of the mth scattering center

Also

$$\frac{1}{\sqrt{2\pi}}\int_{-\infty}^{\infty} h_m(t-T_m)e^{-i\omega t}dt = \frac{1}{\sqrt{2\pi}}\int_{-\infty}^{\infty} h_m(\tau)e^{-i\omega(\tau+T_m)}d\tau$$

$$= e^{-i\omega T_m}\frac{1}{\sqrt{2\pi}}\int_{-\infty}^{\infty} h_m(\tau)e^{-i\omega\tau}d\tau = e^{-i\omega T_m}H_m(\omega).$$

The transfer function can be approximated as an exponential function of frequency [9]. Assuming that $P(\omega)$ is slowly varying then

$$R_E(\omega) = \sum_{m=1}^{M} B_m e^{\tau_m \omega} \qquad (5.36)$$

where

$\tau_m = \alpha_m - iT_m =$ complex times associated with the scattering center impulse

$H_m(\omega) = e^{\tau_m \omega}$

Comparing Eq. (5.33) and Eq. (5.36), we can see a duality between the temporal late-time response and the spectral early-time response.

$$r_{L(t)} = \sum_{n=-N}^{N} A_n e^{S_n t} \qquad (5.33)$$

where $S_n = \sigma_n + i\omega_n$

$$R_E(\omega) = \sum_{m=1}^{M} B_m e^{\tau_m \omega} \qquad (5.36)$$

where $\tau_m = \alpha_m - iT_m$

Now, we want to synthesize an E-pulse in spectral domain ω which will produce a zero output upon convolution with the early-time response:

$$c(\omega) = e(\omega) * R_E(\omega) = \int_0^{\omega_f} R_E(\omega')e(\omega-\omega')d\omega' = 0$$

Taking a Laplace transform,

$$C(s) = L[c(w)] = \int_0^{¥} c(w)e^{-sw}dw$$

$$= E(s)R_E(s) \qquad (5.37)$$

where

$$E(s) = L[e(\omega)] = \int_0^{\infty} e(\omega)e^{-s\omega}d\omega$$

$$R_E(s) = L[R_E(\omega)] = \int_0^{\infty} R_E(\omega)e^{-s\omega}d\omega$$

To have $c(\omega) \to 0$ for $\omega_L < \omega < \omega_F$

It requires $E(s = \tau_m) = 0 \qquad (5.38)$

Then a similar algorithm used in the late-time E-pulse technique can be applied. The demonstration of early-time E-pulse discrimination was reported in details in a paper [10].

REFERENCES

[1] E. J, Rothwell, D. P. Nyquist, K. M. Chen, and B. Drachman, "Radar target discrimination using the extinction-pulse technique," *IEEE Trans. Antennas Propagat.*, vol. AP-33, pp. 929-937, Sept. 1985.

[2] K. M. Chen, D. P. Nyquist, E. J, Rothwell, I. Webb, and B. Drachman, "Radar target discrimination by convolution of radar return with extinction-pulse and single-mode extraction signals," *IEEE Trans. Antennas Propagat.*, vol. AP-34, pp. 896-904, Sept. 1986.

[3] E. J, Rothwell, K. M. Chen, D. P. Nyquist and Weimin Sun, "Frequency domain E-pulse synthesis and target discrimination." *IEEE Trans. Antennas Propagat.*, vol. AP-35, pp. 426-434, April 1987.

[4] C. E. Baum, E. J, Rothwell, K. M. Chen, and D. P. Nyquist, "The singularity expansion method and its application to target identification," *Proc. IEEE*, vol. 79, pp. 1481-1492, Oct. 1991.

[5] E. J, Rothwell, K. M. Chen, and D. P. Nyquist, "Extraction of the natural frequencies of a radar target from a measured response using E-pulse techniques," *IEEE Trans. Antennas Propagat.*, vol. AP-35, pp. 715-720, June 1987.

[6] K. M. Chen, D. P. Nyquist, E. J, Rothwell, and W. M. Sun, "New progress on E/S pulse techniques for noncooperative target recognition," *IEEE Trans. Antennas Propagat.*, vol. 40, pp. 829-833, July 1992.

[7] B. Drachman and E. J, Rothwell, "A continuation method for identification of the natural frequencies of an object using a measured response," *IEEE Trans. Antennas Propagat.*, vol. AP-33, no. 4, pp. 445-450, Apr. 1985.

[8] I. Webb, "Radar target discrimination using K-pulse from a "fast" Prony's method," Ph.D dissertation, Dept. Elec. Eng. Sys. Sci., Michigan State Univ., 1984.

[9] M. P. Hurst and R. Mittra, "Scattering center analysis via Prony's method," *IEEE Trans. Antennas Propagat.*, vol. AP-35, pp. 986-988, Aug. 1987.

[10] E. J, Rothwell, K. M. Chen, D. P. Nyquist, R. Ilavarasan, J. Ross, R. Bebermeyer and Q. Li, "Radar target identification and detection using short EM pulses and the E-pulse technique," Ultra-Wideband Short-pulse Electromagnetic, Edited by H. Bentoni et al, Plenum Press, 1993, pp 475-482.

Index

A

Antenna systems 4-153
- Patch antenna 4-139, 4-141
- Probe antenna 4-139, 4-140, 4-141, 4-151
- Reflector antenna 4-139, 4-141, 4-143, 4-144, 4-146, 4-147, 4-150, 4-151, 4-152

Aspect independent method 5-177
Automatic clutter cancellation circuit 4-123

B

Bessel function 2-44, 2-45, 2-56, 2-61, 2-62, 2-67
Biomedical application 4-117
Boundary condition 1-20, 2-63, 2-73, 2-76, 2-78, 2-79, 2-82

C

Charge density 1-1, 3-97, 3-109, 3-110, 3-112, 3-153, 3-154, 4-156, 4-157, 4-158, 4-159, 4-160, 4-161, 4-163
Clutter 4-123, 4-134, 4-135, 4-137
Correction term 3-96, 3-101, 3-103, 3-109, 3-111, 3-112, 3-115
Creeping wave 2-71
Cross-correlation detection 4-150, 4-151, 4-152
Current density 1-1, 3-97, 3-103, 3-106, 3-109, 3-110, 4-154, 4-161, 4-163, 4-164, 4-167

D

Digital waveform processing sampling oscilloscope (DWPSO) 5-186
Directional coupler 4-123, 4-125, 4-135, 4-136, 4-137
Displacement vector 1-1
Doppler effect 4-118, 4-119
Double-balanced mixer 4-123, 4-124, 4-125, 4-135, 4-137, 4-138
Dyadic Green's function 2-80, 2-81, 2-84, 3-87, 3-91, 3-93, 3-96, 3-97, 3-102, 3-105, 3-109, 3-111, 3-115

E

Earthquake rubble 4-117, 4-133, 4-134, 4-135, 4-136, 4-139, 4-141, 4-142, 4-143, 4-144, 4-145, 4-146, 4-147, 4-148, 4-150, 4-151, 4-152, 4-174

Electric charge 1-2, 1-3, 1-4, 1-20, 4-154, 4-157, 4-158, 4-161

Electric dipole 1-3, 1-4, 3-111, 4-139

Electric field 1-1, 1-2, 1-10, 1-20, 1-24, 1-28, 1-30, 1-31, 1-32, 1-34, 1-35, 1-36, 2-39, 2-54, 2-56, 2-61, 2-62, 2-63, 2-64, 2-68, 2-72, 2-74, 2-80, 2-81, 2-82, 2-83,2-84, 2-85, 2-86, 3-91, 3-92, 3-93, 3-96, 3-97, 3-98, 3-105, 3-106, 3-108, 3-109, 3-111, 3-112, 3-114, 3-115, 4-117, 4-120, 4-122, 4-153, 4-154, 4-155, 4-156, 4-157, 4-160, 4-161, 4-164, 4-165, 4-166, 4-167, 4-168, 4-169, 4-170, 4-171, 4-172, 4-173, 4-174, 4-175

Electric field integral equation 2-72, 2-80, 2-84, 2-85, 4-153, 4-164

Electromagnetics Laboratory of Michigan State University 4-141, 4-142, 5-185

ELF-LF electric field interaction with human body 4-153, 4-175

Equation of continuity 1-1, 3-98, 3-112

Equivalence principle 1-28, 1-31, 1-32, 1-34, 1-36, 1-37, 1-38

Equivalent surface current 1-6, 1-20, 1-23, 1-24, 1-25, 1-28, 1-29, 1-30, 1-31, 1-32, 1-33, 1-36

Extinction-pulse (E-pulse) method 5-177

Extraction signal 5-180, 5-182, 5-184, 5-199, 5-200, 5-201, 5-202, 5-203, 5-210

F

Fast Fourier transform 4-141

Federal Emergency Management Agency 4-134

Fourier transform 1-6, 4-141, 5-207

Frequency-domain analysis 5-195

Frequency-domain signal 4-143

G

Green's function 1-6, 1-10, 1-20, 1-21, 2-80, 2-81, 2-84, 3-87, 3-91, 3-92, 3-93, 3-96, 3-97, 3-102, 3-105, 3-109, 3-111, 3-115,

Green's theorem 1-8, 1-9, 1-28

H

Harmonic equation 2-42

Harrington, R.F. 1-28, 1-38

Health hazard 4-153

Helmholtz equation 2-73, 2-74 , 2-75 , 2-82

Heterogeneous multiple media 1-6, 1-27

I

Infinitesimal current loop 1-2, 1-3, 1-4, 1-5

Irrotational field 2-72, 2-73, 2-80

Induced surface charge 4-153, 4-156, 4-157, 4-160, 4-161, 4-163

Induced current inside the body 4-161, 4-162, 4-169

L

Laplace transform 5-195, 5-209

Legendre polynomial 2-44

Lorentz condition 2-40

M

Magnetic charge 1-2, 1-4, 1-5

Magnetic dipole 1-2, 1-3, 1-4

Magnetic field 1-1, 1-2, 1-4, 1-21, 1-34, 1-36, 1-38, 2-39, 4-156

Magnetic induction 1-1

Magnetization density 1-1

Maxwell's equations 1-1, 1-2, 1-5, 1-6, 1-27, 1-28, 2-80, 4-161

Michigan State University 4-118, 4-134, 4-141, 4-142, 5-178, 5-185

Microwave life-detection systems 4-118, 4-135, 4-141

    L-band life-detection system 4-130

    X-band life-detection system 4-123

Modified Maxwell's equations with magnetic source terms 1-1

Moment method 3-108, 4-158, 4-175

N

Natural frequencies of the target 5-178

Natural modes of target 5-179, 5-181

- Damping coefficient of the nth natural mode 5-179
- Angular frequency of natural mode 5-179

Numerical solution 4-158

P

Permeability 1-1, 1-2, 1-6, 1-28, 1-34, 1-36, 2-72

Permittivity 1-1, 1-2, 1-6, 1-28, 1-34, 1-36, 2-72, 4-120, 4-121, 4-122, 4-156

Phase modulation 4-118, 4-119, 4-120, 4-122, 4-138

Physical optical limit 2-71

Polarization density 1-1

Principal value integration 3-96, 3-99, 3-103

Q

Quarter-wavelength choke 4-140

R

Radar return 5-177, 5-178, 5-179, 5-180, 5-197, 5-201, 5-204, 5-205, 5-210

- Early-time radar return 5-177, 5-205
- Late-time radar return 5-179, 5-205

Radar target identification 5-177, 5-178, 5-195, 5 210

Rayleigh scattering cross section 2-71

S

Scalar potential 3-91

Shadow region 2-71

Single-mode convolved output 5-178

Single-pole double-throw (SPDT) switch 4-136, 4-140

Singularity point 1-10, 1-20, 1-21, 1-23, 1-29, 1-31, 3-99

Spherical Bessel functions 2-44, 2-45, 2-56 , 2-61 , 2-62, 2-67

Spherical scalar wave equation 2-41, 2-57

Spherical vector wave equation 2-46

Spherical vector wave functions 2-46, 2-53 , 2-54, 2-61

  L vector wave function 2-46

  M vector wave function 2-46

  N vector wave function 2-46

Solenoidal field 2-72

Source point 1-14, 3-99, 3-114, 4-157, 4-158, 4-159

Surface-charge integral equation 4-156

T

Time-domain analysis 5-177, 5-178, 5-197

Time-domain signal 4-141 , 4-142, 4-143, 4-146, 4-147, 4-148, 4-151

U

Ultra-wide band signal 5-178

V

Variable attenuator 4-123, 4-124

W

Wave equation 1-8, 2-39, 2-40 , 2-41, 2-46 , 2- 50, 2- 74 , 2-81, 2-82

Z

Zero-mode convolved output 5-178, 5-182, 5-204

Special Topics In Electromagnetics

Publisher	National Taiwan University Press
Author	Kun-Mu Chen
Director	Jieh Hsiang
Executive Editor	Yen-Pei Hsu
Art Design	56MH
Printing	Ching Cherng Print Co. Ltd
Publication Date	Dec. 2008
Edition	First Edition
Price	NT$ 700, US$ 22

Printed in Taiwan, Republic of China

No.1, Section 4, Roosevelt Road, Taipei City 100, Taiwan (R.O.C.)

Tel : 886-2-33663993

Fax :886-2-23636905

http://www.press.ntu.edu.tw

E-mail:ntuprs@ntu.eud.tw

ISBN: 978-986015641-6

GPN: 1009702779